BOEING B-52

BOEING B-52

A DOCUMENTARY HISTORY
WALTER J. BOYNE

Schiffer Military/Aviation History
Atglen, PA

GUARDIANS OF THE UPPER REALM

A B-52H Stratofortress, commonly known as the "BUFF", passes over Mount McKinley on its return to Minot AFB, North Dakota, home of the famous 5th Bombardment Wing. This ship from the 23rd BS is armed with a dozen AGM-86 cruise missiles as a key element in the daily vigilance of Strategic Air Command. Such missions deep within the Arctic Circle forged the 5th BW motto; "Guardians of the upper realm."

The well travelled wing began its combat service with B-17s in the Pacific theater of World War II, and they remained in active service after the war as Strategic Air Command evolved. In 1968. the wing moved to their new home at Minot and received their first B-52G types. They immediately embarked on an extended four-year tour in the Vietnam war, including the "Arc Light" and "Linebacker II" operations.

A decade beyond Vietnam found the 5th BW still in high operational readiness when their B-52H crews won both the Riverside Trophy and Art Neely Trophy as the best of the 15th AF in 1988, as well as taking the Crumm Trophy and Fairchild Trophy as the best unit in SAC overall. They retained both the Riverside and Neely trophies in 1989 for an unprecedented back-to-back victory in those annual events.

Despite the close of the Cold War and the dramatic changes in world politics at the close of the 20th Century, the 5th BW in North Dakota soldiers on. Minot AFB is scheduled to absorb the training operations of the 93rd BW, in 1994 while continuing to fulfill its strategic role. As for the venerable BUFF, the continued alterations of its adaptive platform has extended its service life well into the next century.

Copyright © 1994 by Walter J. Boyne.
Library of Congress Catalog Number: 86-60421

All rights reserved. No part of this work may be reproduced or used in any forms or by any means – graphic, electronic or mechanical, including photocopying or information storage and retrieval systems – without written permission from the copyright holder.

Printed in the United Staes of America.
ISBN: 0-88740-600-9

We are interested in hearing from authors
with book ideas on related topics.

Published by Schiffer Publishing Ltd.
77 Lower Valley Road
Atglen, PA 19310
Please write for a free catalog.
This book may be purchased from the publisher.
Please include $2.95 postage.
Try your bookstore first.

CONTENTS

FOREWORD BY GENERAL R.H. ELLIS, USAF	6
ACKNOWLEDGEMENTS	7
INTRODUCTION	9
THE DEVELOPMENT OF THE AMERICAN HEAVY BOMBER	11
THE INDISPENSABLE FORERUNNER – THE B-47	27
THE DEVELOPMENT OF THE B-52	43
TESTING AND INITIAL PRODUCTION	58
GILDING THE LILY – THE Gs AND Hs	77
THE STRATEGIC AIR COMMAND, FROM BOLLING TO HANOI	89
NEW PLANES FROM OLD	103
MISSILES, BOMBS AND GUNS	123
FLYING THE BUFF	142
APPENDICES	145
INDEX	159

FOREWORD

Conceived in the late '40s, designed in the '50s and last produced in the early '60s, the B-52 Stratofortress continues as one of the world's most formidable weapons delivery systems. The B-52s of the Strategic Air Command have been an essential part of America's nuclear deterrent force structure for well over two decades.

Throughout this long and proud history, B-52 have carried conventional weapons into battle and maintained continuous alert armed with nuclear weapons. They have been equipped, over the years with weapons of ever-increasing technology, such as the Hound Dog air-to-ground missile, the short-range attack missile (SRAM) and the Quail air-launched decoy missile. Today, some models of this extremely versatile manned penetrator are being refitted to carry one of the new weapons of the 1980s—the air launched cruise missile (ALCM). The ALCM, when combined with gravity fall weapons and SRAMS, will enhance the B-52's shoot and penetrate capability.

Although the B-52 was initially designed for high altitude long range strategic missions, armed with nuclear weapons, it has demonstrated its value as a low level penetrator, as well as the ability to carry conventional weapons.

While retaining the option to reconfigure quickly for nuclear missions, the B-52's many unique qualities make it ideal for supporting contingency operations. No other weapon system can match the B-52's capability to deliver concentrated conventional firepower on the battlefield. It also can be used for aerial mine-laying, sea surveillance and reconnaissance, psychological warfare and rapid force projection.

The usefulness of the B-52 has extended far beyond its expected lifespan through research and development of new technologies in weapons and propulsion systems, structural modifications, on-board electronic and computerized equipment, improved flying techniques, and the dedicated efforts of flight and maintenance crews.

The Stratofortress has responded well to the many and varying demands made on it and has earned a lasting position in aviation history.

R. H. Ellis
General, USAF
Commander in Chief
Strategic Air Command

ACKNOWLEDGEMENTS

The support I have received in the process of researching and writing about the legendary bomber, the B-52, has been remarkable. The aircraft inspires affection in the people who have been associated with it, and there was a genuine enthusiasm about turning up exactly correct information, or the precise photo. In the process I was fortunate to meet many capable, intelligent, well motivated and personable individuals who have given a part of themselves to the aircraft—which may explain why the B-52 is not just a weapons system, but something extraordinary, something that speaks for an era.

There is no way for me to put in order the people who made the most significant contributions, for there were so many that helped so much. The United States Air Force was extremely helpful, as was the Boeing Company, as were many others. Perhaps the easiest way is to go through each group, acknowledging the major contributors alphabetically, and then to make a simple listing of all who helped, in the same way. Its not perfect, but there is no perfect way to thank the great numbers of people who assisted me.

In the Air Force I received wonderful support from the Secretary of the Air Force Office of Public Affairs, Magazines and Books Division. There Lt Col Nick Apple (now retired), Sam A. Roberts, III, and Major Paul Kahl were all extremely helpful. In the Defense Audio Visual Agency, fondly known as "Fern Street" to its customers, Dana Bell provided his usual excellent service. I was honoured to have General Richard H. Ellis, Commander of the Strategic Air Command contribute a foreword, and for General Curtis E. LeMay, USAF (Retired) to respond to a series of important questions. The Air Force Museum provided its usual great service, thanks to its Director, Colonel Richard J. Uppstrom, Curator Royal Frey and historians Chuck Worman and Vivian White. Colonel Henry E. "Pete" Warden, USAF (Retired) made an exceptionally helpful contribution, putting many documented facts into proper perspective. I owe a great deal to Brigadier General James R. McCarthy, USAF (Retired) who not only gave me free use of his fine book "Linebacker II: A View From the Rock", but introduced me to Lt Col George B. Allison, who provided many photographs. General McCarthy's book covers the final operations in Southeast Asia, and is highly recommended reading. Among the many public affairs officers who helped, none did more than T/Sgt Harold P. Myers of Castle Air Force Base, California, who went out of his way to provide excellent material. Among individual Air Force contributors, I deeply appreciate the help from Major Dwight A. Moore, Captain Robert P. Jacober, and Captain Mike Mastromichalis.

Within the Boeing Company I must thank first George Weiss, of the Washington Office, Gordon Williams of Seattle, and Al Hobbs and Herb Hollinger of Wichita. They consistently provided good advice, good contacts, good photos and good wishes. L. D. Alford, President of the Boeing Military Aircraft Company in Wichita, gave me complete access to personnel and information at that huge facility. To assist me, Beverly W. "Bev" Hodges sallied forth from retirement to put his finger on people, documents and photos that no-one else could have known existed. "Bev" has a photographic memory and an eye for detail, and he made it possible to do in four exhausting days what couldn't have been done in years without him. Walter House was of great assistance in digging out obscure photos and drawings. Larry Lee and his phenomenal memory helped get a lot of facts into the right sequence. Jack Nelson, whose range of knowledge about the B-52 is amazing, provided me with a briefing he had prepared which was in itself a succinct B-52 history. In Seattle, George C. Martin helped enormously, and it was a privilege to work with him, Edward C. Wells, George S. Schairer, Vaughn L. Blumenthal, Guy M. Townsend and other senior officials of Boeing who were so instrumental in the birth of the B-52. They are all remarkable men who despite their great contributions to aviation are still modest, cooperative, and intensely concerned about engineering. Townsend, a retired USAF Brigadier General, is a great test pilot, raconteur and memory expert, who added many details to the story from his long B-47 and B-52 flying experience, and invested a great deal of time in reading and re-reading the drafts. Dr Paul G. Spitzer and Marilyn Phipps gave me wonderful service from the Boeing Archives—Marilyn often has the material you want before you ask for it—I don't know how she does that. I must also acknowledge the help of the late Harl Brackin, who fought so long to maintain the basis for the Boeing Archival collection. Peter M. Bowers, Vic Seely and Al Lloyd were, as always, generous to a fault and helpful in the extreme.

Among others who helped, I want to thank *Air Force* magazine for putting me in touch with so many B-52 crew members, thanks to John Frisbee. At the National Air and Space Museum, I appreciate the efforts of Don Lopez, who always retrieves my fractured grammar, Jay Spenser and Nancy Harris in reading and re-reading the material. Harvey Lippincott, and his co-worker Bea LaFlamme, provided information and photos on the Pratt & Whitney engines; Jay Miller provided numerous photos and suggestions. I value the efforts of Dom Pisano, the "world's greatest bibliographer", Pete Suthard, Bob van der Linden, Phil Edwards, Dale Hrabak, the "world's greatest photographer", Jim Vineyard and Ed Price. All of these gentlemen on occasion provided me the same excellent, cooperative service that they do to every museum visitor who calls on them.

And, not least, are the following, who contributed facts, photos or good wishes, and to whom I am very grateful indeed. If I have omitted anyone, please forgive me, for my record keeping occasionally got bogged down. From Boeing: Bill Bloxom, Richard M. Curry, Dale Felix, Jack Funk, James H. Goodell, Dick Holloway, Helene K. Little, Marvin Matz, Louise Montell, Don Norby, Earl Norman, E. A. Ochel, Frank J. Parente, Cliff Richards, Will Ryals, John See, Jack Steiner, H. W. Withington. From the Air Force: Phyllis L. Anderson, Captain John Adkinson, S/Sgt David E. Blitch, Colonel John T. Bohn, Art Boykin, Thomas M. Brewer, Gail M. Corrigan, Charles E. Crain Jr., Roger Cummings, James M. Eastman, William Heimdahl, Lt Gerald J. Honeycutt, Major General John W. Houston, William A. "Bill" Jones, Helen Kavanagh, Lt Col. Richard K. Kline, Dave Menard, Lt William Miehe, S/Sgt William Meers, Captain George H. Peck, Jack Reise, Robert Smith, Lt Col Raymond C. Thomas, M/Sgt James O. West, USAF (Retired).

And others: William J. Anders, Gary Bishop, Ron J. Fleishman, Michael Gould, Dr R. T. Jones, Donald N. Hanson, Charles Hansen, Fred Johnsen, Kevin J. Murphy, Robert D. Moreau, Charles Mayer, David R. Nauhey, J. Richard Smith, Marshall T. Slater, Bill Stauffenberg, Ray Towne, Randall B. Thompson and Lawrence Wilson.

Again, if I have omitted anyone, forgive me.

Walter J. Boyne

Washington DC
January 28, 1981

INTRODUCTION

The bomber aircraft was adopted by twentieth-century American military men as readily as the long rifle was by Kentucky frontiersmen. It seemed to be a natural complement first to traditional American policies of isolation by distancing conflict, and later as a means of projecting American interests around the globe after World War II. From the clattering chain-drive Wright biplane from which 1st Lt Roy C. Kirtland dropped tiny hand held bombs in 1910 to the blistering roar of the Rockwell International B-1, bombers have been pre-eminent in the history of US airpower, and none has been so important for so long as the legendary bomber, the Boeing B-52 Stratofortress.

This huge, angular, eight engine aircraft, affectionately called BUFF (for Big Ugly Fat Fellow in the bowdlerized version of the acronym) by its crew members, was designed in story book fashion over a weekend in a Dayton, Ohio, hotel room, and ultimately became a major factor in world politics. The small crowd that on April 15, 1952, watched it lumber for the first time into the air with the strange transformation peculiar to the B-52, as it metamorphosed from a land bound, drooped wing, eight wheeled lumbering freight train into a capable, soaring bird of flight, would not have believed it if they had been told they were witnessing the start of what may well turn out to be a half-century of service.

At the time of writing, the B-52 has already served for 28 years. There are already men who have made an entire military career of servicing the aircraft and there are already young pilots flying it just as their fathers did. Never in the history of modern warfare has a major element of national defence been so long lived. Aircraft were always notorious for how quickly they lost their usefulness. In World War I, designs became obsolete in less than a year; in World War II, the time had lengthened for such classics as the Messerschmitt Bf 109 and the Supermarine Spitfire to perhaps eight or nine years. Even capital ships, battleships and aircraft carriers have had shorter service lives than the B-52.

But longevity does not imply lack of change. The aircraft was improved continuously from the first B-52 on, so that while the external configuration is substantially the same, it is a vastly different machine under its skin.

Three major elements have combined to give the B-52 its amazing life span. The first is the soundness of the original design, which provided a clean, spacious airframe; the second is the amazing cooperation between the Boeing Company and the US Air Force in determining modifications that keep it viable; the third is the Strategic Air Command's crew concept, where the integrity of the crew is maintained through years of training and combat.

The aircraft began as a traditional high altitude heavy bomber, capable of carrying strategic nuclear warheads anywhere in the world. When the development of accurate ground-to-air missiles ruled out incursions at high altitude, the B-52 was adapted for the low level mission, a regime which introduced entirely new factors of fatigue, stress and wear on the aircraft, and required new levels of skill and courage from the crews.

In the Vietnamese conflict, the B-52 was adapted once again, this time to the role of conventional "iron bomb" carrier. What had been conceived as the deliverer of a searing nuclear stilleto to the enemy heartland was now rumbling flying artillery, dropping barrages more reminiscent of the Somme than Hiroshima. The crews, trained to be razor sharp in the precision delivery of a nuclear weapon on an urban target, were just as proficient delivering 750 pound bombs on enemy troop concentrations. They called themselves "coconut knockers" and other names, but they did the job. And when it became time for them to take on the most intensive air defences the world has ever seen, the SAM-studded skies of Hanoi, they performed superbly.

Yet the accomplishments of the B-52 are too often related to its magnificent airframe and the powerful Pratt & Whitney engines which power it, obscuring the massive pyramid of effort which underlies its success. If you analyze these factors, they spread across the country and across the world, in a network of men, women and enterprise that involves hundreds of thousands of people. You can trace this network by beginning with the original sets of principals, the handful of engineers and managers at Boeing who conceived the aircraft, and the handful of leaders in the USAF who spelled out the requirements calling for its conception.

At the next level, we find the hard core engineers who translated these requirements first into drawings, and then, with the talented Seattle team, into metal. Their Air Force counterparts, in Dayton and elsewhere, spelled out the equipment requirements—new and better radar bombing systems, inflight refuelling systems, power generation equipment, electronic countermeasure equipment, more powerful engines and so on. And the industry responded again and again, managing each time to provide the essential new equipment to keep the aircraft serviceable in ever changing environments, and with ever changing missions.

In parallel fashion, workers were trained at Boeing to manufacture the aircraft while airmen were trained in the Air Force to service it; weapons were developed to be dropped from it; pilot training programmes were set up, repair centres established, spare parts lists computed and acquired.

As the aircraft began to leave the production lines for operational service, problems were encountered. Some were fatal, sad and savage accidents relating to equipment failures; others were routine, the ordinary process in a weapon system's development.

The final delivery of the 744th aircraft, a B-52H in October, 1962, was made at a time when the whole industry associated with the plane was in full operation. Thirty-six SAC wings were operating the aircraft at bases all around the US; two factories were equipped for the repair and modification of the fleet, and for the control and manufacture of updating kits. An entire bureaucracy of record keepers meticulously tracked the hours on the fleet, the maintenance that needed to be done, the spares that must be bought.

Yet despite the enormous human back-up, the B-52 has retained its own grudging personality. A monster on the flight line, capable of consuming innumerable maintenance hours for each flight hour, with a repertoire of every mechanical headache from an untraceable fuel leak to complete equipment failures, the B-52s nevertheless always lumber out to meet their take-off times, the wings drooping, brakes squeaking, the cockpits permeated with the smell of JP-4, two decades of indigestible inflight lunches, sweat and smoke. The powerplant runups bring the pounding vibration of eight powerful engines shaking and rattling at their pylon mounts; the skin oilcans, and everything on board, from tool kits to ash trays, vibrates and chatters. The power is brought up, and the aircraft lurches forward, a Disneyesque creature bobbling along at a snail's pace. Soon there is a change, the noises seem to lessen, and the aircraft becomes lighter, faster. The wings, just the wings, begin to fly as the airspeed builds and the runway markers race by. Finally there is that precomputed moment, that balance of weight, temperature, air density, speed and hope where the aircraft lifts from the runway into its own element.

The B-52 doesn't fly like a fighter, nor yet like a truck. It does things its own way, more crisply than one might imagine, but inevitably with the inertia and the space required for 450 000 pounds hurtling at 300 knots indicated airspeed. It is a pilot's aircraft but only in the sense that the pilots that fly it are also engineers, commanders, and when necessary, bronco busters. The B-52 does what it is supposed to do, from flying in close formation to a refuelling Boeing KC-135A tanker, to blistering along low level routes, to specialized tasks like air dropping research aircraft. It does its various jobs well, and will be doing them for a long time.

In this book I will try to set the scene for the B-52 by tracing the development of US bombers, particularly the beautiful Boeing B-47, for which I confess a complete partiality. The B-47 was developed only slightly before the B-52, but its contribution to both civil and military aviation cannot be overestimated. It provided Boeing the jet experience to launch the 707 programme, which in turn has carried the firm to a position of dominance never previously achieved by an aircraft manufacturing company, and it provided the US Air Force with an awesome airpower superiority that achieved total credibility and respect throughout the world, a position we would give much to have today. It gave the Strategic Air Command a degree of capability and sophistication that permitted it to assimilate the much more complex B-52 into its fleet with efficiency and dispatch.

The B-52 story deals not only with the aircraft, but also with the people behind it and with its social and political impact. The aircraft's broad wings have protected the country for more than two decades, and its cultural importance can be seen in everything from news programs to toys to a singing group, "the B-52s"; at the same time it is seen in some countries as the symbol of American aggression.

Over the life of the aircraft past and future, however, there has only been one fundamental purpose, and that is to ensure that it is never used for the role for which it was created—the unleashing of more firepower in a single attack than has ever been delivered in the history of the world. No-one desires this more than the superbly trained crews who fly it. It would be a fitting tribute to an engineering miracle if, when at long last the B-52 finally leaves service, there will no longer be a need for a successor.

THE DEVELOPMENT OF THE AMERICAN HEAVY BOMBER

The heavy bomber has always fascinated the American military mind. The heavy bomber concept was consistent with the way America liked to fight wars—on the other side's turf. If battleships which could keep the enemy from America's native shores are good, then huge aircraft that can fly directly to the enemy's capital and make war there must be even better.

The press has helped this process. No newsman has ever turned down a story about a flying fortress, a leviathan of the skies, a bloody paralyzer of an airplane that can—reportedly—destroy the enemy in a single ferocious fight.

The irony, of course, is that this love affair persisted through 25 years of absolutely minimal performance on the part of the big bombers; it was not until the middle years of World War II that any air force had even part of the capability predicted by the prophets of airpower, Douhet, Mitchell and others.

The 25 years of fame and mediocre ability was a long and painful process of development which saw the gradual evolution of airframes and engines from the original American heavy bomber, the Martin GMB of 1918, to the battle tested B-17s of 1943. The next 25 years were far different, for the ferment and toil and massive expenditure of World War II's aviation development spawned a line of bombers with an almost exponential increase in capability. The capability stemmed in part from the new airframe and engine design which permitted engineering triumphs like the Boeing B-47, but it derived in the main from the introduction of nuclear weaponry, which changed aircraft from what Douhet had hoped for into what he would have feared.

But laborious as the first 25 years were, and as uninspiring as the rate of progress was, there was a thread of human endeavour running through it that makes it fascinating. If each of the painful steps had not been taken, if the development line had halted at any point, the United States would have found World War II to be immensely more costly in blood and treasure, and very probably would have found itself unable to survive in the long cold war which followed VJ day.

Initial American interest in the heavy bomber during World War I had been piqued by foreign designs like the British Handley Page 0/400, and the Italian three engine Caproni Ca 5. Both of these were excellent performers for their day, and both were selected by the Bolling Commission for manufacture in the United States. In the event, difficulties with drawings, measurements and other problems delayed production of the British design by the Standard Aircraft Corporation as planned, although ultimately 100 disassembled and seven assembled planes were delivered to the US Army. The Caproni was beset by similar difficulties, but had the additional problems of an intolerable work arrangement with the Italian representatives in the US. By the end of the war, only three aircraft had been built, two by Standard, and one by the Fisher Body Works of Cleveland.

Far more significant for American aviation history, however, was the work undertaken in 1918 by Glenn L. Martin, a 32-year-old pilot, designer and business genius. Martin had in only ten years brought himself from a few tentative designs, gently cribbed from Curtiss and Wright practice, to the forefront of American aviation. He was able to do what so few of his contemporaries could: make aviation pay. He did this in part by hiring very competent people. He was joined in 1912 by Lawrence D. Bell, the first in a galaxy of bright young men who would collaborate with Martin as long as their nerves could stand it.

The men who worked with Martin, then broke with him, represent major names in American aviation. Bell, Donald W. Douglas, James McDonnell, Dutch Kindleberger and others all paid their dues to the intense young Martin, rebelled against him, and founded their own firms.

Martin sold 17 of his very successful tractor engined biplanes, the TT, to the US Army, and this, coupled with a natural flair for publicity, kept him in sufficient prominence until he could join the Wright-Martin Corporation in 1916, a combine which included the inflated remnants of the Wright firm, from which Orville had long since parted company, the then powerful Simplex automobile company, and various other manufacturers. His new company received permission to manufacture Hispano-Suiza engines under license.

Internal management disagreements led him to leave the firm and form the Glenn L. Martin Company. Impetus for the decision was an order by the US Army for six twin engine, three-place *Corps D'Armée* aircraft with a performance guaranteed to be better than the Handley Page. Martin went to Cleveland, Ohio, armed with ten years of experience, adequate capital, and the best brains in the business. Larry Bell had been with him for five years, and was an excellent factory manager, with an uncanny way with workers, materials and processes. Donald Douglas was his Chief Engineer, a great designer who also had an eye for the requirements of mass production.

On August 27, 1918, eight months to the day from the

Above: The Handley Page 0/400 was built in the United States by the Standard Aircraft Company and powered by two 350 horsepower Liberty engines. Of an initial order for 1,500, seven were delivered assembled, and 100 more were delivered unassembled for shipment overseas. Eventually some of these were assembled in England, and twenty others were stored in the U.S.

Above: The first American heavy bomber, the Martin GMB, seen with four men who would become influential in American aviation. From left, Lawrence Bell, Eric Springer, Glenn L. Martin, and Donald Douglas. The GMB was faster than either the Handley Page or the Italian Caproni Ca. 5 also built in the United States, but had less range and bomb capacity. (U.S. Air Force)

Below: Martin followed up its success with 20 of the more advanced MB-2 aircraft, which were subsequently redesignated NBS-1. When a further 110 of the type were to be ordered, the three low bid contractors proved to be Curtiss, L.W.F. and Aeromarine, and Martin was left without a product. This is a Curtiss NBS-1, one of 50 the firm built, which was slower than the GMB, despite its cleaner lines, but carried more bombs farther. (U.S. Air Force)

receipt of the order, a beautiful twin engine biplane, the Glenn Martin Bomber, rolled out of the Cleveland factory doors. Eric Springer, an experienced pilot, made the first flight and pronounced it "the best job I ever flew".

An order for 50 of the aircraft was placed on October 22, and it seemed that there at last would be an American warplane which could "darken the skies over Europe". Optimistic talk of the production of as many as 1,500 of the GMBs was bruited about until November 11, 1918.

The Armistice did not quite mean the end of the line, although the initial order for 50 was slashed to ten. Fortunately for Martin, these ten performed so well, and the need for pride in an American product was so great, that they provided the basis for the company's future success.

Perhaps Martin's greatest achievement was that the GMB did in some ways exceed the performance of the world famous Handley Page. The Martin was slightly smaller, but it had a top speed of 105 mph compared to only 92 for the version of the 0/400 which was equipped with Liberty engines. Both aircraft had similar service ceilings of about 10,000 feet, but the British design had the greater bomb load and a longer range.

The Glenn Martin Bombers, retrospectively called MB-1s, had a checkered career, being used mostly for testing of a variety of weapons and equipment. Twenty of an improved version, the MB-2, were offered and these were the aircraft used in the battleship bombing tests off the Virginia Cape. General Billy Mitchell had sufficient foresight not only to secure the Martins for the tests, but to pressure the arsenal at Aberdeen, Maryland, to create 1,000 and 2,000 pound bombs, whose near-misses sent the *Ostfriesland* to the bottom.

Thus the stage had been set for ten years of American bomber development. Even in the fund starved 1920s, the US Army Air Service was able to buy 110 MB-2s, which became the backbone of the American bomber force for the next five years.

The most convincing argument for the importance of the Martins is a look at the aircraft which tried to succeed it. The Elias XNBS-3, the Curtiss XNBS-4, the Huff-Daland XLB-5 and the entire line of more than 200 Keystones all faithfully followed the Martin formula of biplane wings, twin engines and lots of drag.

There were some departures from the formula, usually with results that ranged from the freakish to the grotesque. The huge three engine LWF Owl, the earlier J. V. Martin "Cruising Bomber", the implacably slow but incredibly famous six engine Barling bomber, and the impossible Johns Multiplane are good examples. An isolated success was the Curtiss B-2 Condor, of which only 12 were built because of the high procurement cost of $82,000. The contemporary Keystones were smaller and sold for less than $30,000, and the Chief of the Air Corps, Brigadier General J. E. Fechet, preferred quantity to quality.

The Keystones formed the bulk of Air Corps bomber strength for almost seven years, with very little to recommend them but their low purchase price. They were ponderous on the controls, and capable only of stretching a glide on a single engine. A good quality, discovered by more than one pilot, was that you could walk away from virtually any crash landing if you could just stick the wing in the ground first, and let the yards of steel tubing, fabric and wire absorb the shock.

Keystone did not have the capacity to build metal aircraft, even though it proposed an angular, bulky prototype to the Air Corps, and when in 1931 the Boeing XB-9 "Death Angel" appeared, it was clear that the long line of Martin GMB inspired bombers was at an end.

In the B-9, biplane wings had been replaced by a single thick cantilever structure that housed well cowled engines and retractable landing gear. Construction was all metal monocoque, just as had been used on its predecessor, the

Below: Competitors tried to improve on the basic Curtiss/Martin design, but did not succeed. This is the Elias XNBS-3, which had a steel tube fuselage and wooden wings which did not pass the quality control tests of the U.S. Army Air Service inspector. (U.S. Air Force)

Below: *The desire for something different sometimes led designers to the bizarre. This is the J.V. Martin (not related to Glenn L. Martin Company) Cruising Bomber, which was not accepted for service. It had inboard engines driving propellers through a none too strong transmission system. Next to it is the equally unsuccessful J.V. Martin Kitten, which featured retractable landing gear, and still may be seen at the National Air and Space Museum's Paul E. Garber Facility. (U.S. Air Force)*

Left: *Curtiss tried a larger version of the formula, with the XNBS-4; with a 90 foot wing span and 435 horsepower Liberty engines, it offered no significant advance.* (U.S. Air Force)

Above: *Much more successful, in relative terms, was the rather handsome L.W.F. "Owl", which featured advanced monocoque fuselage construction and the usual drag inducing wing, empennage and gear arrangement of the period. It is shown in flight over Hazelhurst Field, New York. Speed was 110 mph, and range with 2,000 pounds of bombs was 1,100 miles, though it was never tested to its capacity.* (Courtesy Col. Phillips Melville)

Left: *The Johns Multiplane shows the optimistic extremes to which inventors went during the immediate postwar period. It was damaged on its only attempt at flight.* (U.S. Air Force)

Left: *"Bloody Paralyzing"* rather than a bloody paralyzer, the Barling Bomber was none the less the darling of the press, and was strongly advocated by Captain Billy Mitchell. It actually set several records, with General Harold R. Harris at the controls, but despite its six Liberty engines could reach only 96 mph in speed, and had a pathetic 170 mile range with a load of bombs.

Centre left: *Giving up the single Packard installation used on their postwar XLB-1 and XHB-1, Huff-Daland (shortly to become Keystone) reverted to the twin Liberty formula, and eventually arrived at a successful series. This is the LB-5A, of which 25 were built. Performance was still, in 1927, almost identical to World War 1 standards, with a top speed of 107 mph, range of 435 miles and a 2300 pound bomb load.* (U.S. Air Force)

Bottom left: *The Curtiss B-2, designed by George Page, was a really excellent aircraft for the time, with a top speed of 132 mph, 780 mile range and 2,500 pound bomb load. Unfortunately it was too expensive for the limited budgets of the time, and only twelve were purchased. It led to the Condor passenger aircraft.* (U.S. Air Force)

Top right: *Boeing, having apparently secured a firm place in the fighter market, built upon the success of its Monomail to create the XB-901 (YB-9) in 1931. A service test model, the Y1B-9 shown here, had cleanly cowled Pratt & Whitney R-1860-11 radial engines, retractable landing gear and a top speed of 188 mph, faster than the biplane fighters of the day. It is shown here with its equally revolutionary fighter counterpart, the Y1P-26.* (U.S. Air Force)

Right: *The Martin XB-907 "Mystery Ship" saw active duty as the Martin B-10, a very advanced aircraft for its day, and one which at last freed the Air Corps from its bondage to biplanes. The B-10 made several notable flights, including one Hap Arnold led to Alaska and back, but its principal value was in acquainting Air Corps leaders with real performance and modern maintenance requirements.* (U.S. Air Force)

17

Monomail, and these features gave the angular but handsome aircraft an enormous jump in performance. Two prototypes were built on speculation, one the Model 214 with 575 horsepower Pratt & Whitney R-1860-13 Hornet radial engines, and the other, the Model 215, powered by 600 horsepower Curtiss V-1570-29 Conqueror liquid-cooled inline engines. The radial engines were preferred, and seven aircraft (including the two prototypes) were purchased for service test for $692,324.

Boeing was convinced it had a winning design, for the B-9 had a top speed of 188 mph, faster than contemporary biplane fighters. For the first time, a twin engine bomber had a reasonable single engine capability, so that the danger from engine failure was reduced rather than multiplied by the use of two engines.

Unfortunately for Boeing, the Air Corps engineers at Wright Field had not had the smoothest of relationships so far with the Martin company. Wright Field, still wearing the mantle of its pioneering predecessor, McCook Field, had been attempting to elicit a modern bomber from Martin with its design directive MN-24 of early 1929. Martin was amazingly unresponsive, attempting first to offer a conventional fixed gear biplane, and as a second choice, an ungainly fixed gear monoplane. Martin officially recommended the biplane "because of the advantages inherent in the smaller overall dimensions" but the real reason was probably their greater familiarity with orthodox construction.

There ensued an endless series of conferences, memoes, and impassioned long-distance telephone calls (a really dramatic managerial device in the 1930s) which proved to be a prelude to a tempestuous two year marriage of convenience. When this had ended, the Material Division at Wright Field was exhausted, but had the aircraft it wanted. Martin was delighted, for it had built the XB-907—dubbed the "Flying Whale" or "Mystery Bomber", depending upon the newspaper you read at the time.

The Martin company had also won the continued priviledge of interacting with Wright Field to produce a really modern bomber. Wright Field insisted upon a number of modern concepts which included the use of a special wing construction which had the ribs and box spars covered by

Below: *The American fascination with size was exemplified by the 149 foot span Boeing XB-15, which proved to be a very practical aircraft in eight years of service. Underpowered, it was primarily a lesson in the technique of building very large aircraft. First flown in 1937, the XB-15 is shown here with a Boeing P-26.* (U.S. Air Force)

span wise sections of corrugated metal, which were in turn covered by a smooth outer skin; the use of large fillets to smooth out the airflow over the stabilizer; a significantly stronger, cantilever empennage, and an improved single strut gear retraction system. The latter had been devised by Jean A. Roche, doyen of the McCook Field/Wright Field engineering group, and father of the famous Aeronca C-2 lightplane.

As the project matured, a host of other improvements were incorporated, including new engines, more streamlined cowlings, a revolutionary glazed revolving turret for the forward machine-gunner, and after much experimentation, stylish sliding canopies over the pilot and rear gunner compartments.

Martin gained a great deal from the long involvement, for besides winning the Collier Trophy in 1932, it ultimately sold 154 of the aircraft to the Air Corps and an additional 189 to foreign buyers.

The new airplane permitted serious consideration of new tactics and new equipment which would enable the Air Corps to break out of the rigid mould of World War I style operations. The Martin B-10 paved the way for strategic operations in the true sense, for its advent coincided with the development of the highly accurate Norden bombsight. The combination led to the formulation of the precision bombing philosophy which dominated the thinking of US air leaders ever after.

The prospect of a weapon which could fulfil Douhet's theories was intoxicating to Air Corps leaders, and thoughts began to turn to much bigger projects. The late Brigadier General Benjamin Kelsey, USAF (Retired), Lindbergh Scholar at the National Air and Space Museum, tells of this in his book "The Dragon's Teeth?". In brief, the B-10 validated the concept of the truly powerful long range bomber in the minds of such ardent proponents as Hugh Knerr, Frank Andrews, Hap Arnold and Tooey Spaatz.

Ironically, the uncomplimentary press reports which had attended the Air Corps' ill-fated assignment to carry the mail in 1934 had resulted in government financial appropriations larger than ever before in peace-time. The bomber men were now able to experiment on a bigger than ever scale and their

Below: *One step up in scale was the Douglas XB-19, with its 212 foot wing span, and maximum gross weight of 162,000 pounds. Powered by Wright R-3350 engines, the aircraft had a top speed of only 224 mph, but was well liked by its pilots. It too was primarily an exercise in learning to build giant aircraft.* (Douglas)

Above: *The four engine Boeing entry was a typical Seattle response; it was determined to give its customer what it needed rather than what it had asked for. The four engine interpretation of the "Multi-engine" formula was acceptable for test purposes, but Boeing had run the risk of being disqualified. The prototype Model 299's performance was as revolutionary as its appearance.* (Boeing)

Right: *The prototype Fortress crashed on October 30, 1935, through pilot error.* (U.S. Air Force)

point of departure was the 74 foot wing span, 14,731 pound gross weight Martin B-10. Calculations for a bomber with a range of 5,000 miles and a 2,000 pound bomb load resulted in an aircraft more than twice the B-10's size. The first was the Boeing XB-15, with a 149 foot wing span and a 70,700 pound gross weight. The next step, providing the same range with a 4,000 pound bomb load, proved to be the Douglas XB-19, a 212 foot wing span, 160,000 pound giant.

Both aircraft were underpowered, and their performance relatively poor, but each proved the practicality of building very large aircraft, and cleared the conceptual way for the postwar Consolidated-Vultee B-36, and ultimately the Boeing B-52.

The design, study and development of the XB-15 gave Boeing confidence to enter a four engine design in a competition for a "Multi-engine bomber". The other entrants, an advanced model of the B-10, the Martin Model 146, and the Douglas DB-1, had clung to the twin engine formula, and were completely outclassed by the Boeing Model 299, the prototype of the enormous fleet of B-17s which followed. It was an early example of Boeing's habit of giving the customer what it thinks it needs instead of exactly what it asked for.

The model 299 (which because of a fatal accident never became the XB-17 except in retrospective error) flew on July 28, 1935, its 103 foot wing span dwarfing all of the other aircraft at Boeing Field, Seattle. Powered by four 750 horsepower Pratt & Whitney Hornet engines, the aircraft captured public fancy, its streamlined blister gun positions (which later proved to be totally ineffective) immediately inspiring the name "Flying Fortress".

The true potential of the silver giant was revealed to the military when it flew 2,100 miles from Seattle to Wright Field on August 20, 1935, in a record setting time of just over nine hours, averaging a 233 mph cruising speed for the trip. The fastest contemporary Air Corps fighter, the Boeing P-26 "Peashooter" had a top speed of only 234 mph.

The Model 299 had been brought into being by a dedicated team of managers, engineers and workers who realized that their future livelihood depended upon the success of the aircraft. There was fostered a sense of commitment and integrity which permeates the firm to this day, despite its remarkable growth in size.

The original design team had been relatively small. Headed by Claire Egtvedt, who had joined Boeing in 1917 as an assistant engineer, it had E. Gifford Emery as Project Engineer and an eager 24-year-old Edward C. Wells as Assistant Project Engineer. All three men had distinguished careers at Boeing, particularly Wells, who would play a very important role with the B-52.

The small group laboured at a design whose modern appearance was matched by a host of features which set the pace for bomber design. The innovations included controllable pitch constant speed propellers, automatic mixture controls, an efficient cockpit layout and a favourable power to weight ratio.

The original planform did not change significantly from the Model 299 to the last B-17G. Control surfaces were enlarged, bigger, turbosupercharged engines were em-

Above: *The Flying Fortress became literally that over Hitler's Festung Europa,* as ever more formidable armament and equipment was added. This B-17G had a maximum takeoff weight of 65,500 pounds, carried thirteen .50 caliber guns, and could dispose of three to five tons of bombs. 8,680 G models were built. (Boeing)

ployed, self-sealing tanks, armour and specialized combat equipment were fitted, but the basic design did not change.

Fate intervened when the prototype Model 299 crashed on take-off at Wright Field; Major Ployer Hill had taken off with the control lock on, an accident similar to the one which claimed the life of the Chief of Staff of the *Luftwaffe, Generalleutnant* Walter Wever in June 1936, in a Heinkel He 70. The 299 stalled, dived into the earth, and killed both Hill and the Boeing Chief Test Pilot, Les Tower.

The crash tragically seemed to suit the Congressional style of approving the purchase of the most aircraft, not necessarily the best, with the available appropriation. Only 13 service test Y1B-17s were purchased, along with 133 of the Douglas B-18s, as the DB-1s were redesignated. The B-18 had no growth potential; it was only a mediocre performer at best and would serve out the war as an anti-submarine aircraft and as squadron hacks.

The crash almost destroyed Boeing financially. Somehow the firm managed to survive on small orders for 39 B-17Bs in 1938 and 38 B-17Cs in 1939. By then, however, the European war was so imminent that production orders began to increase. Britain received 20 of the Boeing B-17Cs on Lend-Lease, calling them Fortress Is, but their combat career was inauspicious, tending to confirm the already tightly held British view that night area bombing was the proper strategy.

The sale of first line warplanes to France and Britain had a *quid pro quo* for the United States in that the Allies were to supply technical information on equipment required by combat conditions. As a result, later B-17s were equipped with self-sealing fuel tanks, heavier armament, heavier armour, and so on. Thus the B-17Ds that went into combat in late 1941 and early 1942 were not totally helpless, even if they were not completely successful as warplanes.

The miracle of the 1935 Model 299 is that it was able to digest an endless stream of changes engendered by the war and become in time the really capable B-17 Model Fs and Gs which ultimately hammered Germany into the dust.

These later aircraft, which flew in huge box formations day by day over Germany, were enormously successful. They were able to accommodate the higher gross weights caused by additional fuel and equipment and still have formidable performance. This inherent ability to "grow" became a fundamental aim of Boeing designers, and manifested itself in extraordinary fashion in the Boeing B-52.

Pilots loved the B-17; it was equally popular with ground crews. The stories of its rugged construction surviving battle damage are legion, and it was to soldier on, in a variety of civil guises, for more than 30 years.

Ultimately 12,731 Flying Fortresses were built, including license built models by Lockheed and Douglas. The aircraft established Boeing as the dominant American bomber builder.

At the risk of offending the many fans of the Consolidated B-24 Liberator, its story will be condensed inasmuch as it did not bear on the ultimate design of the B-52.

A stablemate of the B-17 throughout the war, the B-24 did not have to go through the harrowing gestation period of its rival. The Air Corps in 1939 wanted an even better plane than the B-17 and specified a top speed of more than 300 mph, a range of 3,000 miles and a ceiling of 35,000 feet. The B-24 never quite achieved these ambitious goals, but it came close, and in the end was built in greater quantity (more than 18,000) than any other US aircraft.

Never as popular with the public as the Flying Fortress, the Liberator was actually much more versatile, serving in a variety of passenger carrying, anti-submarine patrol and freighter roles, as well as achieving real success as a bomber in both the European and Pacific theatres of war. It had features which clearly showed it to be of a later generation than the B-17. Its tricycle landing gear, twin rudders, boxy

Above: *The most significant bomber of World War II, the Boeing B-29 is shown next to its predecessor, the B-17. The B-29 was the first aircraft to be able to combine outstanding performance with a nuclear weapon, and thus alter military planning for all time.* (U.S. Air Force)

fuselage and high aspect ratio Davis wing did not endow it with the B-17's handsome appearance, nor did its more sophisticated systems endear it to ground crews. While many of its pilots were highly partisan, praising the aircraft and claiming to pity the poor Fortress drivers, the fact that the USAAF elected to eliminate the Liberators from service almost immediately after the war speaks volumes for the relative regard in which the two planes were held.

The Next Generation

Even before the B-17Gs began coming down the production line, the second giant step toward the B-52 was under test. The XB-29, the Superfortress, had been ordered into development in early 1940 as a "Hemisphere Defense Weapon", a hedge against the real possibility that European bases might not be available to the United States.

The XB-29 was a formidable gamble, even though it represented continuous development by Boeing of the big bomber idea through a series of design studies which had

Left: *Standby for the B-29 was the Consolidated B-32 "Dominator". What would have been an unqualified success in any other country was relegated to a minor role. This is the first XB-32, with twin tails; later versions had the single tail similar to the Privateer in outline.* (General Dynamics)

Centre left: *Turboprops were seen as a possible powerplant for the B-29; unfortunately, the engines never materialized in time.* (Boeing)

Below: *The B-29 was vastly improved by the addition of the much more powerful Pratt & Whitney R-4360 "Corncob" engines of 3,500 horsepower. Distinguished by their modified nacelles and tall tails, the B-50's welcome additional horsepower was accented by the much more reliable P&W engines. The use of an engine analyzer in B-50s made diagnosis of incipient engine problems much easier, and resulted in a vast saving of maintenance hours as well as a big improvement in safety.* (Boeing)

The Northrop XB-35 had survived as much on novelty and promise as anything else. The Air Force was reluctant to cease experimentation with something that looked so good on paper. In the event, the airplane, which had an outstanding performance, was converted to the jet-powered YB-49. (US Air Force)

Above: *The Convair B-36 represented the fulfillment of the dream of the intercontinental bomber. No one who has ever heard one go over can forget the rumbling roar of stately power. With six P&W R-4360s and four J47-GE-19 jets, the B-36 "Peacemaker" had a range of 7,500 miles with 10,000 pounds of bombs, and a top speed of 440 mph at 32,000 feet, with everything at maximum power.* (Convair)

followed the XB-15. While the B-17 had stemmed directly from previous Boeing practice in terms of aerofoil, structure and general design philosophy, the B-29 was a complete departure, one that could only have been taken by a company increasingly confident in its own capability.

The B-29's structural innovations were in fact the design philosophy upon which the later Boeing jets were built. Boeing abandoned its traditional aluminium alloy tube built-up wing, with its characteristically deep aerofoil and relatively low wing loading, for more modern web construction, a low drag aerofoil and a reduced wing area which resulted in an exceptionally heavy wing loading for the time. It was this basic concept, including the use of huge Fowler flaps, from which the even thinner and more sophisticated B-47 wing evolved.

Similarly the B-29 fuselage was much more advanced than that of any previous Seattle product. Using aluminium of much higher tensile strength, and a cylindrically shaped fuselage which housed pressurized crew compartments, the B-29 was immensely strong, as befitted the heaviest US aircraft to enter production in World War II.

There were other gambles, too, the greatest and most costly being the selection of the trouble prone 2,200 horsepower Wright R-3350 Cyclone engines. These engines were ultimately developed to a useful state, but for most of the B-29's career they plagued it with internal fires and other mechanical defects. Other refinements in the production aircraft included remotely controlled General Electric gun turrets, the APQ-7 radar bombing system, nose wheel steering, double bomb bays, a central fire control system and more.

Despite all the unknown factors, the great need of the US Army Air Force was matched by its confidence in the Boeing company, and in 1941, long before the September 21, 1942 first flight of the prototype, more than 1,660 of the gigantic aircraft had been ordered. This was a significant allocation of America's production potential, and the failure of the design would have had enormous repercussions.

The B-29 did not fail. Rushed from drawing board into operational use within four years, it proceeded to lay waste the cities and industries of Japan. Three billion dollars and an incredible mobilization of industrial strength produced a war winning weapon, one which at last confirmed all the theories of the airpower advocates. Its first combat sortie against Japan took place on June 15, 1944; one year later most of Japan's principal cities were no longer worthwhile targets, having been burned into ashes by Major-General Curtis E. LeMay's Twentieth Air Force.

The B-29 had won the war in the Pacific unleashing attacks whose ferocity made December 7, 1941 seem insignificant. It then capped its own efforts with two missions, the single aircraft atomic devastations of Hiroshima and Nagasaki. The great fleets of B-29s which flew over the battleship *Missouri* during the September 15, 1945 surrender ceremonies were far more than a tribute to Douhet, Mitchell, Knerr, Andrews, Spaatz and Arnold; they were a signal to the future.

Despite the fact that the B-29 and its refined piston engine successor, the B-50, were to soldier on for another 20 years in various roles, the silver B-29s were already obsolete, condemned by new shapes on the Seattle drawing boards. The jet engine which the British and Germans had brought into use in Europe was already being contemplated for bombers.

THE INDISPENSABLE FORERUNNER – THE B-47

The solid experience of building the B-29s resulted in a plane which brought joy to the hearts of bomber pilots. After years of listening to garrulous fighter pilots spieling endlessly of their alert and responsive mounts, of enduring their witless bar-room jibes about aerial truck drivers, bomber pilots suddenly had in the B-47 a plane that was almost fighter-like in its flying qualities.

It was a pleasure to fly the slim six jet B-47. Despite the myths that surround it concerning its unforgiving ways, its requirement for airspeeds to be flown with impossible precision, its inevitable flight to the "coffin corner" where high speed stall and low speed stall were just knots apart, the B-47 was a lovely aircraft to fly, particularly for veterans of the B-29 and B-50.

The sleek lines of the aircraft gave a first impression of speed and liveliness that familiarity did not dim. The fighter-like cockpit, with its blown canopy and tandem seating added to the illusion. But most important was the fact that the B-47 was responsive in almost all regimes of flight. At very heavy weights, on hot days, at high altitudes, it was sluggish on take-off, and runways did tend to seem very short. But even under these conditions, once airborne the B-47 assumed delightful characteristics.

The speed was exhilarating; for someone brought up on four flailing piston engines, the climb speed of the B-47, starting at 310 knots indicated airspeed, was a heady experience. The rate of climb was fantastic, and the high cruising speeds, normally about Mach .74, or 425 mph true airspeed, made cross country flight a pleasure. Russ Schleeh, then a Major, and Major Joseph W. Howell set an unofficial transcontinental record on February 8, 1949, when they flew the prototype XB-47 from Moses Lake, Washington, to Andrews Air Force Base, Maryland in three hours, 46 minutes. It was Schleeh's second flight in the airplane, and Howell's first.

The visibility from the cockpit was panoramic, and at altitude, in the early days of the B-47, you were the only ones there. If you saw a contrail, it was another Stratojet.

Tactical operations were a pleasure. The radar observer in the three man crew was kept busy all the time, sharing duties with the aircraft commander and co-pilot. Bomb runs, navigation legs, let downs, all became precision tasks in which the crew coordinated and cooperated.

Inflight refuelling was, at high gross weights, somewhat dicey when the tanker was the piston engined Boeing KC-97. These tankers, perfectly suitable for B-50s, often had to refuel in a full power descent in order to stay above the B-47s stall speed, which naturally increased as it took on fuel. The entire refuelling process with the KC-97 was inefficient, because the B-47 used almost as much fuel to descend to refuelling altitude, rendezvous, refuel, and come back to cruising altitude as it was able to take on from the tanker. When the turbine powered KC-135A tankers came into general service, refuelling became a much more productive task.

The B-47 was the first of a totally new generation of aircraft where airframes rather than powerplants became the limiting factor on speed. Much has been written about the critical requirement to maintain exact airspeeds in the descent and landing phases of B-47 operations. The requirement was there, but so was the capability to fulfil it; a caress of the throttles would result in a minute flickering of the percentage calibrated RPM counters, and airspeed could be controlled exactly as desired. In a crosswind, a slight differential between the number one and number six engines would keep you right on track during an ILS (instrument landing) approach.

The B-47 was flown in formation on occasions, usually for show purposes, and at lower altitudes it was relatively easy to keep in position. At higher altitudes, however, the aircraft was so "clean" that station keeping was more difficult. The author will never forget a 12 ship formation from the 93rd Bomb Wing, formed up in four elements of threes, at about 25,000 feet over the Sierra Nevadas. A new hot shot major had just joined the 330th Bomb Squadron, and he was hurrying to catch up with the formation. He came barrelling in out of the west, and, fighter style, whipped his B-47 up on one wing to allow aerodynamic braking to slow him down and allow him to join up. Even in what seemed to be a 60 degree bank, his B-47 didn't slow down a knot, and he sailed through the formation in knife edge flight, missing the second element lead by just a few feet, and sailing on out toward Nevada with a unanimous stream of curses following. He joined up a few minutes later in a more conservative manner, and was pretty quiet around the squadron for a few days.

There were losses with the B-47, in the early days primarily due to the rapid roll due to yaw of the swept wing. If an outboard engine failed just as you became airborne, the pilot had three seconds to apply sufficient rudder forces (15 degrees!) to maintain directional control. This was a particularly hard lesson to learn, for the normal tendency in a jet is to minimize use of the rudder, and pick up a wing with aileron.

Later in its service there were metal fatigue and other problems which caused a series of accidents.

When the B-47 fleet was at its peak in the mid-1950s and early 1960s, the United States enjoyed a measure of strategic superiority greater than any nation had enjoyed before—or since. In those halcyon days of the Strategic Air Command, the B-47 was virtually immune to attack. At altitude it was almost as fast as the North American F-86s, Grumman Panthers and other fighters that vainly tried to intercept it, and it would not have had much of a problem with MiG-15s. With more than 2,000 B-47s built, and perhaps as many as 1,500 ready to launch in the event of a war, the United States could have destroyed any aggressor in a matter of days, with relatively few losses. And it is only fair to note that this awesome power, this incredible fleet of virtually invulnerable aircraft, was not used.

As time went on, and the century series fighters and their foreign counterparts came into service, the B-47's measure of superiority faded. After years of orbiting pre-selected sites, and talking fighters in for an intercept, one bright sunny morning a North American F-100 Super Sabre came from nowhere and barrel-rolled around us. We knew then that the B-47's heydey was over.

Development and Test

The B-47s began with the sudden awareness that the Germans and British had startling new powerplants in production, and were building both jet fighters and jet bombers. The jet fighter spelled the end of the piston engine bomber just as surely as the ironclad spelled the end of the wooden man of war. The US Army Air Force began calling for jet bomber design studies as early as 1943, and by April, 1944, was specifying a large aircraft with a 500 mph top speed. Five designs were tendered, ranging from conservative to radical. North American offered the straight wing XB-45, which became the first US jet bomber to go into production, and which served operationally as the Tornado for several accident filled years. Convair developed the most elegant appearing of the lot, straight winged also, but possessing an almost feminine beauty. It was the XB-46, but despite its magnificent looks and adequate performance there was never a chance that it would go into production for the firm already had a large contract for B-36s and, as in all postwar periods, the Air Force's policy was to spread the funding. (The B-36, called the Peacemaker, would lumber along on its six piston and four jet engines for years, flying missions of interminable length. The B-36 was the pinnacle of the past in bomber development, while its partner in peace, the B-47, was the foundation stone for the future.)

Below: *The Douglas XB-43 was the first American jet bomber. A conversion of the static test XB-42 "Mixmaster", the "bug-eyed" bomber was overtaken by the swiftly advancing technology and did not go into production, but instead served as an excellent test bed at Edwards Air Force Base for many years under the nickname "Versatile II". It is currently in the National Air and Space Museum's Garber Facility.* (Warren Bodie)

Above: *The North American XB-45 "Tornado" was the first American jet bomber to go into production. Of conventional design except for the new powerplants, B-45s served the USAF from 1947 through 1958, seeing combat in the Korean conflict. Top speed was 579 mph. (U.S. Air Force)*

Martin offered the least attractive design, the boxy XB-48, which shared only one thing with the B-47, its bicycle type landing gear. Martin had insisted on using a large cowling for each of its two sets of three engines, and these created enormous drag.

Northrop adapted its XB-35 flying wing to the eight jet YB-49 configuration, and radical as the design was, it offered excellent performance.

All of these aircraft flew in 1947, beginning with the XB-45 on March 17, the XB-46 on April 2, the XB-48 on June 14 and the YB-49 on October 21. When the Boeing XB-47 first flew on December 17, 1947, it was evident that a radical new era had dawned, and that Boeing had no real competition.

The XB-47's configuration—long slender fuselage, narrow, high aspect ratio swept back wings, six engines arranged in underslung pods, and a bicycle landing gear—was arrived at only through a long and sometimes anguished process at the Boeing design desks.

The first Boeing jet design, the Model 424, looked like little more than a B-29 fitted with four jet engines paired in nacelles under the wing. One version followed another from the Boeing drafting boards, as more was learned about the problems associated with jet flight, problems which included high fuel consumption, the danger of fire resulting from failure of the all too frangible turbine blades of the day, and the effect of drag that conventional aerofoils and wings of the day had at the high speeds which the jet engine promised.

Model 432 retained a straight wing and had the engines disported about the fuselage centre section, right over the fuel tank area, with bulbous air intakes located besides the cockpit section. Model 432 was an almost classical expression of the 1944 dilemma facing aero engineers. The L/D (Lift to Drag) ratio of straight wings fell off drastically at Mach .6 because of the vastly increased drag due to Mach effect; on the other hand, at speeds below Mach .6, jet engines were only about two-thirds as efficient as contemporary reciprocating engines. One solution, tentatively essayed in Germany, favoured for civil use in England, held in second class status in the US and vastly laboured over in Russia, was the turboprop. The turbine engine/propeller combination seemed to be the only feasible solution to the drag/efficiency equation.

Robert T. Jones, of NACA, believed that a better solution to the problem was to use swept back wings. His paper on the

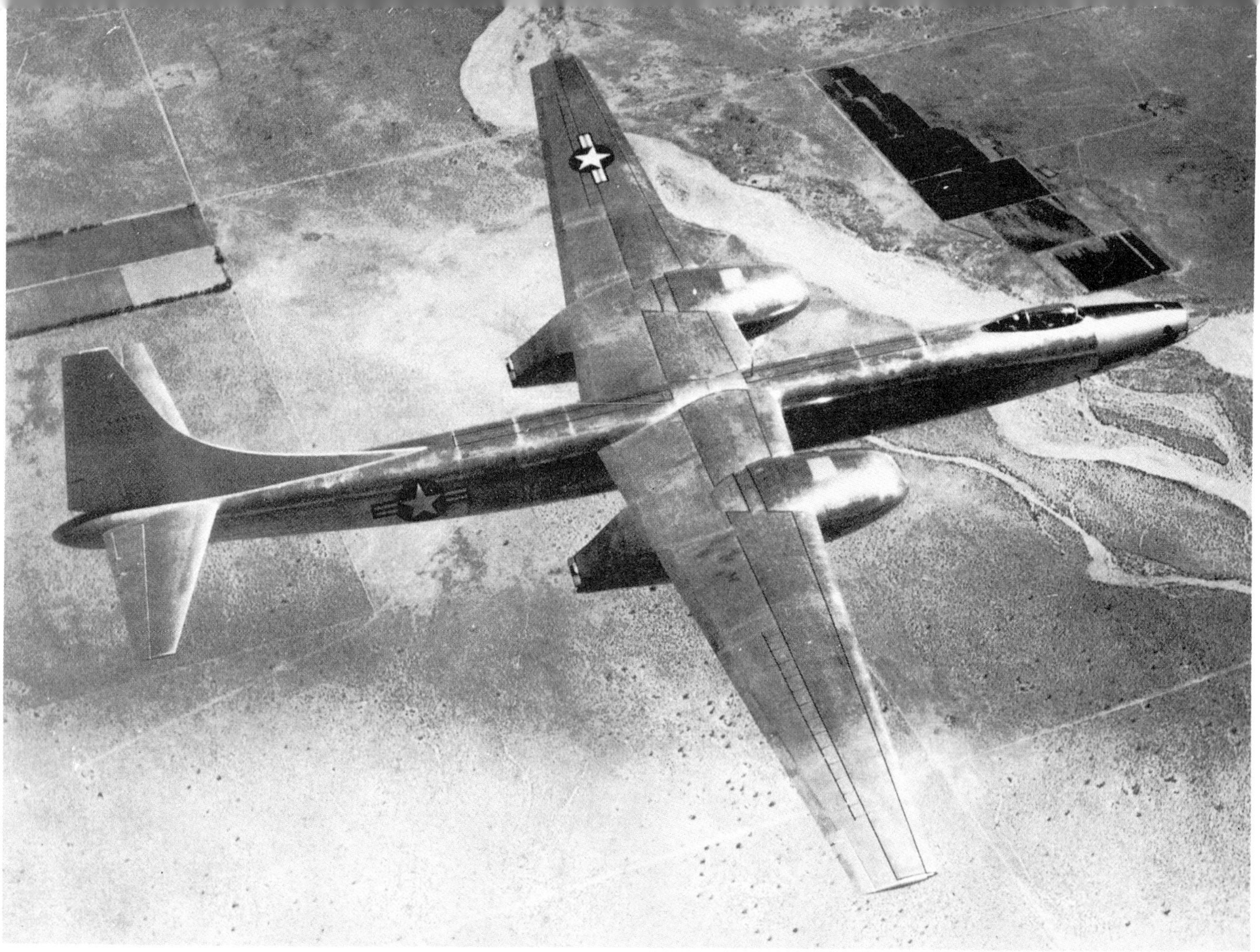

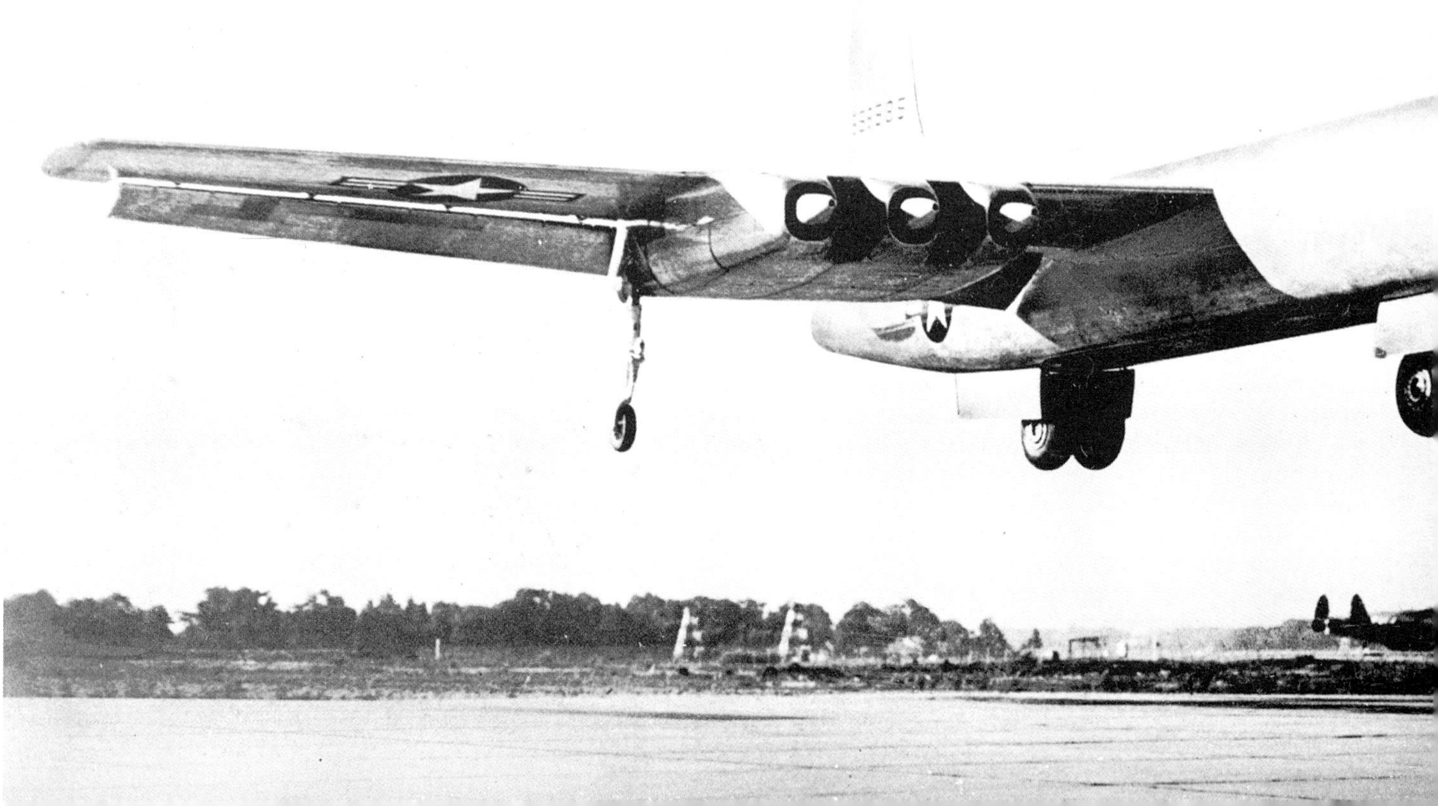

Left: *The most beautiful of the group of experimental jet bombers, the Convair XB-46 did not have a chance of going into production because the firm was already a supplier of B-36s. Its cockpit design was considered so excellent that it was sent to Wright Field as a standard for jet bombers. Top speed was about 490 mph. (Convair Photo)*

Below: *Martin's XB-48 was a somewhat ungainly design; the huge nacelles, each housing three engines, created formidable drag. The Martin designed bicycle landing gear was adapted for the rival Boeing B-47. (Martin)*

Right: *Most radical of all the designs was the Northrop YB-49, essentially the XB-35 re-engined with eight Allison J35 engines. Performance was excellent, and there has been a constant hint of hard-ball politics being connected with its demise, Northrop reputedly having been told to merge with General Dynamics in order to reduce the number of companies competing for contracts. (U.S. Air Force)*

Below right: *The group of men most responsible for the Boeing B-47. From left, George Schairer, Edward C. Wells, George Martin, and Robert "Bob" Jewett. Even these men had no idea they were laying the basis for a multi-billion dollar military and commercial fleet of swept wing jets. (Boeing)*

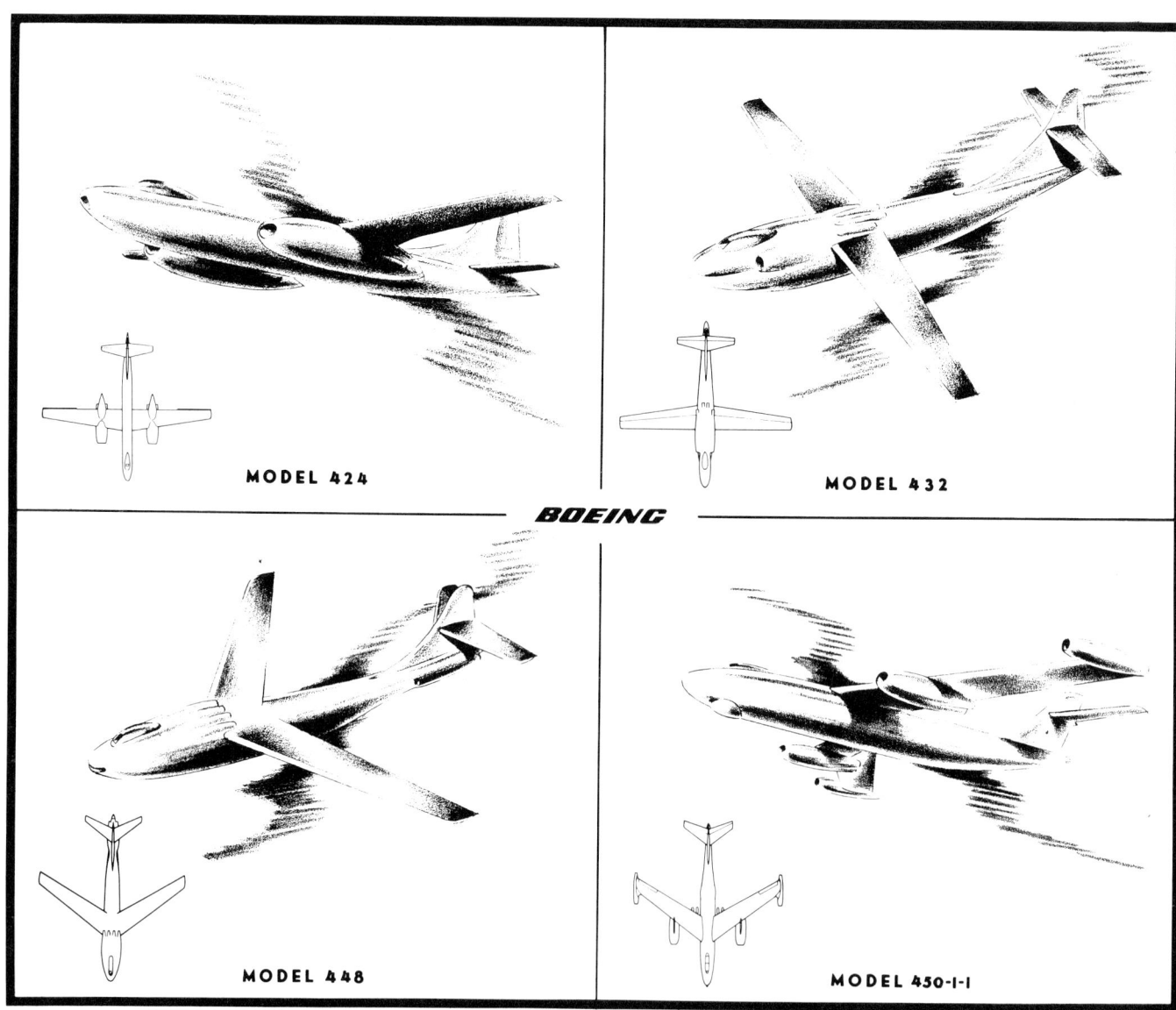

MODEL 424

MODEL 432

MODEL 448

MODEL 450-1-1

subject was turned down for publication on the grounds that he lacked experimental proof, even though swept wing aircraft were already flying in Europe. (The Messerschmitt Me262, which had 18½ degrees of wing sweep and was the first operational jet fighter, had its wings swept for structural and centre of gravity reasons rather than as an attempt to delay the critical Mach.)

In the closing days of World War II, the US Army's Scientific Advisory Group, under Theodore von Karman, went to Europe to find out the latest German aeronautical developments. George Schairer, Boeing's Chief Aerodynamicist, accompanied the group which included such luminaries as Frank Wattendorf, Hsu Shen Tsien, Hugh Dryden, and others. They found in German records sufficient material to validate Jones' theory. Interviews with German engineers, including Adolf Buseman, who had first proposed the theory of swept wings in Vienna in 1935, caused Schairer to write back to Seattle and recommend the adoption of sweepback.

As a result, the next model in the progression towards the

Above: *The B-47 design passed from the conventional Model 424 which was almost a B-29 with jet engines, through the unlovely Model 432 with its bulbous air intakes and dangerously positioned jet engines, through the similarly rotund but swept wing Model 448 to the close-to-final Model 450-1-1.* (Boeing)

Above right: *Innovations of the B-47 included high aspect ratio, thin wing, pod mounted engines, bicycle landing gear, small crew, built in "Jato" rockets. Oddly enough, the B-47 project was relatively low budget, calling for use of existing systems throughout in order to concentrate on jet engine and swept wing configuration development.* (Boeing)

Right: *The aircraft which finally resulted, the XB-47, was a classic design, clearly superior to all of its competitors, and destined to revolutionize both bomber and transport aircraft. Drag was far less on the aircraft than anticipated, resulting in a potential for a really long range bomber—the B-52.* (Boeing)

B-47, Model 448, retained the rotund fuselage of the Model 432, but featured swept back wings. The very thinness of the high aspect ratio swept back wings caused other problems. The thin wing did not allow fuel, engines or landing gear to be stowed in it, and these problems were solved in the Model 450-1-1, which looked almost like the final B-47 configuration. The inboard engines were suspended in pods to avoid interference with the clean wing, and both fuel and landing gear were concentrated in the fuselage.

The pods were one of those happy engineering circumstances where better aerodynamics also meant better structure. The weight distribution of the engines along the span permitted the wing to be made lighter. As a side effect, they also made maintenance easier, and in the event of fire or damage, were less likely to extend the condition to the wing itself. As a final unforeseen advantage, the inboard nacelles helped to induce a stall over the section of the wing next to them, while the outboard tip mounted nacelles tended to delay the stall. This permitted greater aileron effectiveness near the stall, but more importantly, delayed pitch-up, the nemesis of swept wing aircraft of the time.

A mock-up inspection was held in Seattle in April 1946, and the Air Force was monumentally impressed. Boeing was given the go-ahead, in May 1946, with two prototypes being ordered. The mock-up resulted in the wing span being increased to 116 feet, and the outer engines being also placed in pods. The Martin-originated "Stump-jumper" bicycle undercarriage was adopted.

This radical aircraft, so unlike any previous bomber, made its first flight 19 months after go-ahead with Bob Robbins and Scott Osler at the controls. In retrospect, it is probable that not even the Air Force fully understood the weapon it had been given, just as Boeing could not have realized that the basic cornerstone of a many-billion dollar bomber and commercial jet transport business had been laid.

George Martin was Project Engineer on the B-47, and he insisted that the programme be focused almost entirely on the two major challenges, that of the new swept wing aerodynamics, and on the jet engine itself. Working with a relatively limited budget, Martin adapted some readily available components for use in the B-47 programme, including elements of the B-29 landing gear, a low pressure hydraulic system, and an obsolescent 28 volt DC electrical system.

Even so, the B-47 was precedent setting. It was not a perfect weapon, for the early jet engines, even at their relatively low thrust output, consumed fuel in enormous quantities. Solutions to the range problem were perceived in inflight refuelling, and eventually the B-47 would routinely fly bone-shattering 24 hour missions. While its speed was phenomenal for the time, its altitude capability was not startling, but at lighter "over the target" gross weights, it was certainly respectable. Most of all, it was ready for marriage to the remarkable new bombing and navigation systems which were coming in to being, and with the fast developing US atomic arsenal.

Structurally, the aircraft was far more sophisticated than anything Boeing had built. The leap from the tubular aluminium alloy bridge truss construction of the B-17 to the webbed wing technique used in the B-29 was great, but the transition to the thick-skinned B-47 was greater. The 116 foot wing span, with its 9.43 to 1 aspect ratio, the highest to date of any jet bomber, was extraordinarily flexible, and capable of a deflection of 17½ feet, tip to tip, in static tests. It was also flexible across the chord, and at speeds above 425 knots indicated the ailerons acted as a tab, twisting the wing rather than causing the ailerons to assist in a bank. At 456 knots the ailerons became absolutely ineffective and you were devoid of lateral control. In the normal operating envelope, however, there was plenty of lateral control at all speeds.

As so often happens in aircraft design, solutions to one problem created others. The bicycle gear, adopted because there was no room to stow the undercarriage in the wings, dictated that the aircraft take-off and land at a fixed attitude, rather than with the customary rotation or flare of conventional aircraft. The gear arrangement also required some new techniques for taxiing, but pilots quickly adapted to it.

The huge Fowler flaps chosen for the B-47 were designed to give maximum lift and low drag. Because the early jet engines took a long time to "spool-up" for acceleration, it was advisable to keep power on during the landing approach, so that an emergency climb and go-around could be made if necessary. "Carrying" power in the super clean B-47, even with full flaps, made it very difficult for the pilot to put it down exactly on the end of the runway, where he had to be, in view of the relatively short runways of the time. The solution to this was the approach chute, which permitted the pilot to keep the engines in an RPM range where he could rapidly accelerate for a go-around, but still provide enough drag to permit spot landings. Spoilers were considered as a possible answer to this problem, but a trial installation on the prototype revealed that they had disastrous impact on the stall speed. A brake chute and an anti-skid braking system helped get the aircraft stopped once on the runway.

With only a three man crew, it was possible to group them all together in a small pressurized compartment. The radar observer (radar observer/navigator/bombardier) sat in a dark cubby hole unrelieved by windows in later models, while the two pilots had an unrestricted field of view.

Despite its formidable performance, the Air Force did not continue to evidence a great deal of interest in the aircraft, for it seemed to fall midway between the requirements for medium and heavy bombers. General K. B. Wolfe, who had masterminded the B-29 programme, and was continually urging Boeing on in the B-50 programme, made a perfunctory trip to Moses Lake, Washington, the XB-47 test site.

Major (later Brigadier General) Guy M. Townsend had taken over the test programme, and Colonel Henry E. "Pete" Warden, of Wright Field, and who would be perhaps the single most important Air Force figure in the B-52 programme, wanted Wolfe to take a trial flight. Wolfe wasn't interested, but Warden kept at him for almost 45 minutes before he finally agreed to make a short flight. Townsend made a spectacular take-off, climbed at a speed which made Wolfe inquire if the instruments were reading accurately, and then let him fly it. When they had landed, Wolfe called Warden over and said "You'd better come see me about this." From then on the Air Force was fully behind the B-47.

The original XB-47 had been powered by 3,750 pound

static thrust Allison J35-2 engines, which were so primitive that they actually used farm machinery bearings, and bearing temperatures became a critical factor in their use. The total of 21,500 pounds of thrust by six of the J35s was marginal even for the 125,000 pound maximum gross weight of the prototype. The second prototype had six General Electric J47-GE-3 engines each with a thrust of 5,200 pounds static, which proved more than adequate. These engines were also retro-fitted to the number one prototype.

The development process passed swiftly and soon led to a contract for ten B-47A aircraft, dated September 3, 1948. These were essentially test aircraft, as important for the lessons they taught about production of a sophisticated jet bomber as for their test results. They were fitted with the J47-GE-11 engines of 5,200 pounds static thrust.

The B-47A was built at Boeing's Wichita facility, which had turned out as many as four B-29s a day during the war. Located in the heart of Beech and Cessna country, and on the same site as McConnell Air Force Base where B-47 training would take place, the plant responded to the challenge of the new aircraft in spite of atrocious weather conditions. No-one who was there can forget the numbing Kansas winters, with the wind whistling mournfully, or the excruciatingly hot dry summers, where cockpit temperatures sometimes exceeded 160 degrees Fahrenheit. The McConnell ramp would be a noxious ocean of unburned J-4 fumes, waves of mirage inducing heat, sticky asphalt, and sweating crew members. After hours of searing pre-flight activities, there was almost miraculous relief after take-off when the aircraft's air conditioning system would bang into action, filling the cockpit

Above: *Perhaps the most influential Air Force personality associated with the B-47 programme was Major Guy M. Townsend (later Brigadier General), an absolutely charming individual with a steel trap mind; he combines the best qualities of determined test pilot and level headed engineer. Townsend would be equally important in the B-52 programme. (Boeing)*

Below: *Early jets depended upon refuelling capability to provide required combat range. B-47s and KC-97s were only just compatible; at higher gross weights, KC-97s had to refuel in a full-power descent to keep airspeed above the B-47's stalling speed. The Boeing designed boom system was a decisive advance in aerial refueling techniques. (U.S. Air Force)*

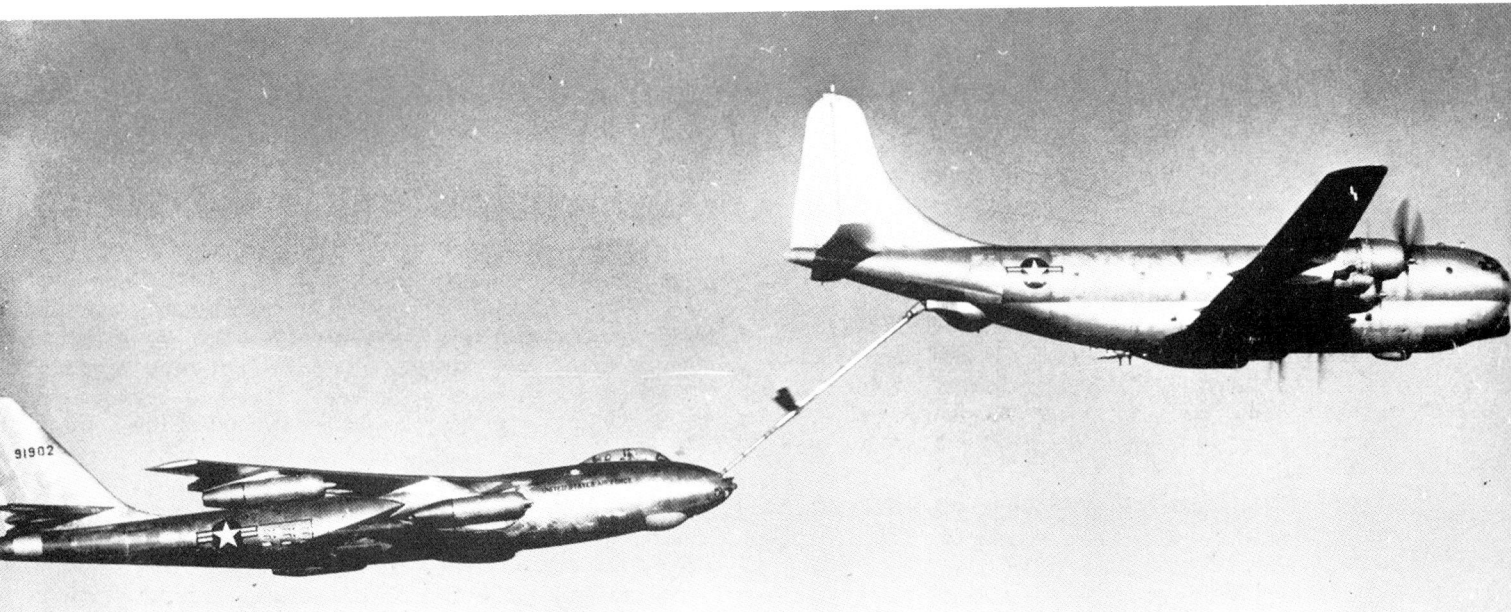

with a chill fog that looked unnervingly like smoke. It was better than a cold beer, although you had to calm anyone on his first flight, for he was usually ready to abandon ship because of the "fire".

There were many historical forces at work which determined the extent of the B-47 programme. The Berlin Airlift (June 24, 1948 to May 12, 1949) had called world wide attention to the escalating cold war, and there followed a series of similar crises which shocked the US Congress into appropriating enormous sums of money to buy the swept wing bombers.

The lessons of the A model having been digested, 87 B-47Bs were ordered in November 1948, and this aircraft featured the absolutely indispensable inflight refuelling capability. Boeing had solved the low fuel rate delivery problem of the existing probe and drogue systems with its innovative flying boom, and incidentally sewed up future sales for KC-97 and KC-135A tankers.

The B model was greatly improved in other ways also, including a structural beef-up and a change in load limit factor from 3 to 2 which permitted the all-up weight to go to 200,000 pounds. It still did not have ejection seats, however, a dreadful defect that made SAC crews nervous for two years before the B model was turned over for training purposes.

A total of 399 B models were ordered, and these not only provided the basis for training and tactics, but also underwent structural and development tests for the later programmes. They also provided the entry for Lockheed and Douglas in the sudden acceleration of B-47 procurement. Lockheed built eight B-47Bs and Douglas ten, the basis for their later full-scale participation in the B-47E programme.

The B-47C and D models were experimental only. The B-47C was a paper study for a four engine B-47 with much greater range; it was, as we shall see, a threat to the B-52 concept and was dropped on General LeMay's request. Two B-47Bs were converted to XB-47Ds to serve as test beds for a turboprop installation. A Wright YT49-W-1 turboprop was placed in lieu of the two J47s in each inner nacelle; the outer nacelle retained the J47 jet engine. The aircraft was fast, faster than any other propeller powered aircraft at 597 mph, but engine out conditions rendered it virtually uncontrollable, and the project was dropped.

The B-47E was the definitive Stratojet. General Electric J47-GE-25 engines of 6,000 pounds thrust were augmented to 7,200 pounds by water injection, and new rocket assisted take-off racks were fitted consisting of 33 1,000 pound thrust units in a jettisonable horseshoe collar fitting. Tail armament was improved with the addition of a General Electric radar directed rear turret with two M-39 20 mm cannon. The co-pilot's seat would swivel 180 degrees, and he could pick up at a limited range and azimuth any enemy fighter obliging enough to make a lengthy stern attack. The guns worked well enough in training; their use in combat would have been problematic.

Orders continued to flow for B-47s, so much that all three manufacturers began essentially wartime production efforts on the B-47E. Before production concluded, Boeing had built 931, Douglas 264, and Lockheed 386 of the E model, and SAC had the most formidable unchallenged strike force in history.

Operational Use

The B-47, as impressive as it was, would not have been effective if the Strategic Air Command had not made the same quantum jump in operational efficiency that the aircraft did in aerodynamic efficiency. SAC did not come into existence as an organic entity, a fully developed continuation

Below: *The speed of the jet bomber required ejection seats. In this B-47B-II the two pilots eject upward while the radar observer (as he was called in those days) ejected downward. Some radar observers used to remind their pilots, somewhat plaintively, that in the event of an ejection on take-off that they should roll the aircraft somewhat to give the RO a fighting chance.* (Boeing)

of the vast air armadas of World War II. Even the officers who witnessed the decline of American airpower from 1945 to 1948 could scarcely believe the change in attitude and in operations. There was in this interim period a lack of organization, a softness, totally uncharacteristic of the US Army Air Force of World War II. In part it had to do with the fact that the war had been won, and the threat of the Soviet Union had not yet been fully realized. In part it had to do with the massive outpouring of qualified personnel to civil life; in part it had to do with the vastly reduced budgets of the period.

There were other elements too, not easily identified, and perhaps not found everywhere in SAC. They were present in 1953 when the author joined the 330th Bomb Squadron of the 93rd Bomb Castle Air Force Base, California, an idealistic young 2nd Lieutenant just graduated from the Aviation Cadet programme.*

The 330th flew B-50Ds, the refined "super B-29s" that filled the gap before the jet bombers were ready. The squadron was peopled with experienced types from World War II, great personalities, but with a playful attitude towards their mission. The Operations room was used for roll call and flight briefings until about 9:00 AM; following that it was used by all hands, officers and enlisted, to play hearts for the rest of the day. On my first gunnery mission, as a green co-pilot, I watched the gunners load ammunition on the bomb bay doors. I hesitantly asked the lead gunner if it wouldn't be easier to load the guns on the ground, rather than having to go into the bomb bay during flight. "Don't you worry about it, Lootenant", he said.

I found out why; over the gunnery range, the bomb bay doors were opened, the ammunition, still in boxes, jettisoned, and the gunners scored 100% without having to clean their weapons afterward.

There were lots of other abuses; bombardiers regularly used the Norden bomb sight to give their radar operators a hand on radar bomb drops; pilots would mash down on the tone signal which was used on the simulated bombing runs to interfere with signals from the runs of other aircraft from other squadrons. It wasn't so much corruption as a light hearted flippancy and laziness.

Mercifully it all ended one Monday when a team from SAC Headquarters swept in and began firing squadron commanders and shifting crews about. A completely new standard of ethics was instituted overnight, and with it a vast improvement in technique, morale, and performance.

The intensive process which SAC used to endow individual crews with a great sense of loyalty, dedication and professionalism was to pay dividends in Viet Nam.

There were other, more visible signs of SAC's managerial and operational efficiency. It grew almost tenfold between its formation in 1946 and 1959, when it reached a peak in terms of numbers of personnel and aircraft assigned. At the same time, the safety record improved enormously, and all measures of performance—navigational accuracy, bombing accuracy, gunnery, etc—were improved enormously. Best of all, perhaps, there was a continuous improvement in the flying safety records.

The B-47 programme flourished in these years. By 1952 there were four wings equipped with 45 B-47s each; by 1953 there were seven; by 1954, 17; by 1955, 22; and by 1956, 27. The peak of 28 wings was reached in 1958, with four additional wings of RB-47s.

SAC could muster 1,367 B-47s in 1958, along with 176 RB-47s and 380 B-52s. These were operated by highly trained crews, who could fly as much as they needed to in order to maintain superlative standards in radar bombing, inflight refuelling and the hundred other tasks by which Headquarters, and this meant LeMay, evaluated combat efficiency. Fuel prices were low, and operational use was limited more by maintenance and crew factors than anything else.

The onus for performance was placed squarely on the unit commanders. If an enlisted man got drunk and had an accident, it was assumed that his squadron commander was directly responsible, and there was a rocket duly delivered.

These glowing years of American airpower were studded with individual feats which grabbed headlines and characterized the business-like growth in efficiency. The first operational B-47 was delivered from Wichita to MacDill Air Force Base on October 23, 1951, to the 306th Bomb Wing. Colonel Michael N. W. McCoy made the flight; he would later be killed in a crash resulting from aileron ineffectiveness in a high speed turn. Pinecastle Air Force base would be renamed for him.

In 1953 the 306th initiated a five year programme of rotating B-47 wings through British bases where the presence of American nuclear power would be even more visible to the Russians. From June 1953 through April 1958, one or more B-47 wings was on station in the United Kingdom.

The 22nd Bomb Wing, whose roots run to the bloody combat of the South Pacific in 1942, appropriately made the first flight of B-47s to the Far East. On June 21, 1954, three B-47s flew non-stop from March Air Force Base, California, to Yokata Air Base, Japan, in 15 hours, covering the 6,700 miles with two inflight refuellings. This flight was alluded to in the Jimmy Stewart film "Strategic Air Command". Later in the same year a combination of bad weather and recognized opportunity induced Colonel David A. Burchinal, Commander of the 43rd Bomb Wing, to establish a distance and endurance record in the B-47, flying 47 hours and 35 minutes and covering a distance of 21,163 miles, back and forth between RAF Fairford and Sidi Slimane, French Morocco.

The demonstrated success of inflight refuelling and the prospect of the introduction of the turbine powered KC-135 tanker fleet permitted a revision of SAC's planning. Instead of deploying to overseas bases and conducting strikes from there, B-47s would in the future make their strikes directly from the United States, and either return to the Continental US if sufficient refuelling capability was available, or else make a post-strike landing at a friendly foreign air base.

The entire achievement of the SAC force was based on the

---

* The Aviation Cadet programme permitted young men without a college degree to enter flying training. These men were highly motivated, and wanted to fly and have an Air Force career more than anything else in life. The decision to drop the Aviation Cadet programme and insist on a college degree as a prerequisite for flight training is in the minds of many a costly mistake, for a college graduate often has other options than an Air Force career.

Above: *The Boeing B-47E was the definitive version of the aircraft, bringing SAC's offensive power to a new high. Powered by General Electric J47-GE-25 engines, which with water injection had a thrust of 7,200 pounds, the E model was a formidable performer. Thirty three RATO units, mounted in an external horse-collar, helped on heavyweight takeoffs.* (Boeing)

Below: *A beautiful aircraft. Thin white line immediately in front of canopy is a yaw string, a primitive bit of instrumentation which worked perfectly.* (Boeing)

long hours of intensive work put in by every level of personnel. All over the country, an 80 hour work week was nothing to comment on; if it was required, it was done. There were curious anomalies in the system as it grew. When hand tools were in such short supply, crew chiefs shared them, one set between each two B-47s. It was crazy to have two $3,000,000 aircraft sitting side by side with a $400 tool kit between them, and two master sergeants arguing over whose turn it was to use a ratchet wrench, but that's the way it was. Tools came out of one pocket, and aircraft out of another.

LeMay made his personality felt down to the newest airman; he was symbolically looking over your shoulder at all times, and often enough he was actually there, looking. The *esprit* was magnificent; the morale numbing ground alert system had not yet been initiated, and the full capabilities of the B-47 were just being proven, so everyone had a feeling of direct participation.

The high morale was reflected in a number of ways. Crews began sprucing up; instead of the sometimes rancid flying suits of the past, clean flying clothes, with natty coloured scarves, became the order of the day. One affectation that many pursued was a LeMay-like cigar jutting out from the jaw; no matter that it made you sick, or that it made the cockpit smell bad, if LeMay did it, it was bound to be good.

The requirement for mobility and flexibility was uppermost in SAC's planning, and entire combat wings and air refuelling squadrons would rotate to Britain, Africa or Alaska, Enormous amounts of thought were placed into what "fly-away" kits would carry, so that maintenance could be carried on with a minimum of interruptions. Similarly, new concepts in maintenance resulted in changes in inspection methods and times, overhauls, inventory systems and so on. The SAC budget, while large, was finite, and the operation of more than 3,000 tactical aircraft called for great quantities of fuel, spares and even such ordinary commodities as parkas, sun glasses, boots, dog food for K-9 corps watch dogs and so on.

All of this effort reached an operational peak with two simulated combat missions during December, 1956. More than 1,000 B-47s and KC-97s flew missions all over the Northern Hemisphere; in many respects it was the contemporary war plan applied to the West. More than anything else it proved that with the B-47, SAC was able to execute its mission with a precision never before achieved by a peacetime air force, and with more potential firepower than had been expended in all the wars of previous history.

This was the peak in America's relative military advantage which after 1957 was steadily eroded by the Soviet missile build up. SAC strength would increase, but not at as rapid a rate as the Russians. The B-52 strength was rising; by 1962, only ten years after its first flight, there would be 639 Stratofortresses available. The B-47 phase out programme incredibly had begun in 1957, the same year that the 100th Bomb Wing became the last to equip with the B-47. The phase out stemmed from a number of factors, including the limited life expectancy of the B-47 airframes, the increased capability of the B-52 units, and not least, the requirement to change operational techniques so as to be able to have one-third of the force on ground alert. This strategy, in which one-third of the SAC fleet was on alert at all times, ready to take off in 15 minutes, was required because Soviet ICBMS could now target all of the SAC airfields.

Once the B-47 phase out began, it continued with terrifying rapidity. The number of B-47 bombers fell from their 1958 peak of 1,367 to 880 in 1962, 391 in 1964 and 114 in 1965, the last year of their operational use. They soldiered on for a year in the EB-47 electronic ferret role, and then had a variety of test and other duties, but their once invincible numbers were soon reduced to aluminium ingots at the Davis-Monthan Air Force Base disposal site. One of the saddest sights ever seen by an old B-47 pilot was row upon row of beautiful aircraft being guillotined to pieces and then melted down.

Operational Problems

The B-47 was an effective weapon system, but it is only fair to mention that the aircraft encountered some very severe problems in the course of its service.

Many of the problems stemmed from the fact that it was a radical design, operating at speeds and altitudes that no heavy bomber had been capable of before. The original 125,000 pound gross weight had been increased steadily to 230,000 pounds for taxi, and a maximum inflight weight of 225,958 pounds.

The B-47 was limited to two positive gs at maximum gross weight, 1.5 gs with flaps down, and no negative (a proper double negative, for once) gs. With the great speed of the aircraft, its power controls, and the capability to enter fighter-like banks, the g limits were doubtless often exceeded. In addition, some pilots could not refrain from rolling the aircraft, although all aerobatics were strictly prohibited.

Not surprisingly, the aircraft had shown some routine fatigue problems early in its operational career. One involved replacement of the fuselage-wing drag angles, and the other the replacement of some panels which cracked near the outboard engine. New tactics, required to offset the increasing threat from ground-to-air missiles, imposed entirely new strains on the aircraft, however, and catastrophic problems began to occur.

The tactics involved two separate manoeuvres. One, called "pop-up", required the bomber to fly low, just off the ground, then pull rapidly up to 18,000 feet to release its weapon before turning violently away and dropping back down to ground level. The second, called LABS, for low altitude bombing system, involved the aircraft roaring in on the deck, pulling up in a half loop, releasing the bomb at the top of the manoeuvre, then rolling out in an Immelmann turn to provide clearance from the impending nuclear explosion.

It was not an exceptionally difficult manoeuvre to do, except that the aircraft was large, bomber pilots were not used to aerobatics, and it was easy to exceed g limitations.

Much worse, however, was the fact that the manoeuvres induced fatigue, and there were a series of failures in 1958 which rocked SAC just as failures in the Vickers Valiant would rock the RAF in 1964. Between March 13 and April 15, 1958, six B-47s crashed, and these triggered a fleet wide investigation. The inspection revealed widespread fatigue problems, ranging from fatigue in the lower wing skin at buttock line 45,

XB-47D was a conversion from standard B by installing Wright YT49-W-1 turboprop engines in lieu of inboard jets. Although it turned out to be the fastest propeller driven aircraft ever, the engines were not satisfactory. Assymetric power control characteristics were abysmal. (Boeing)

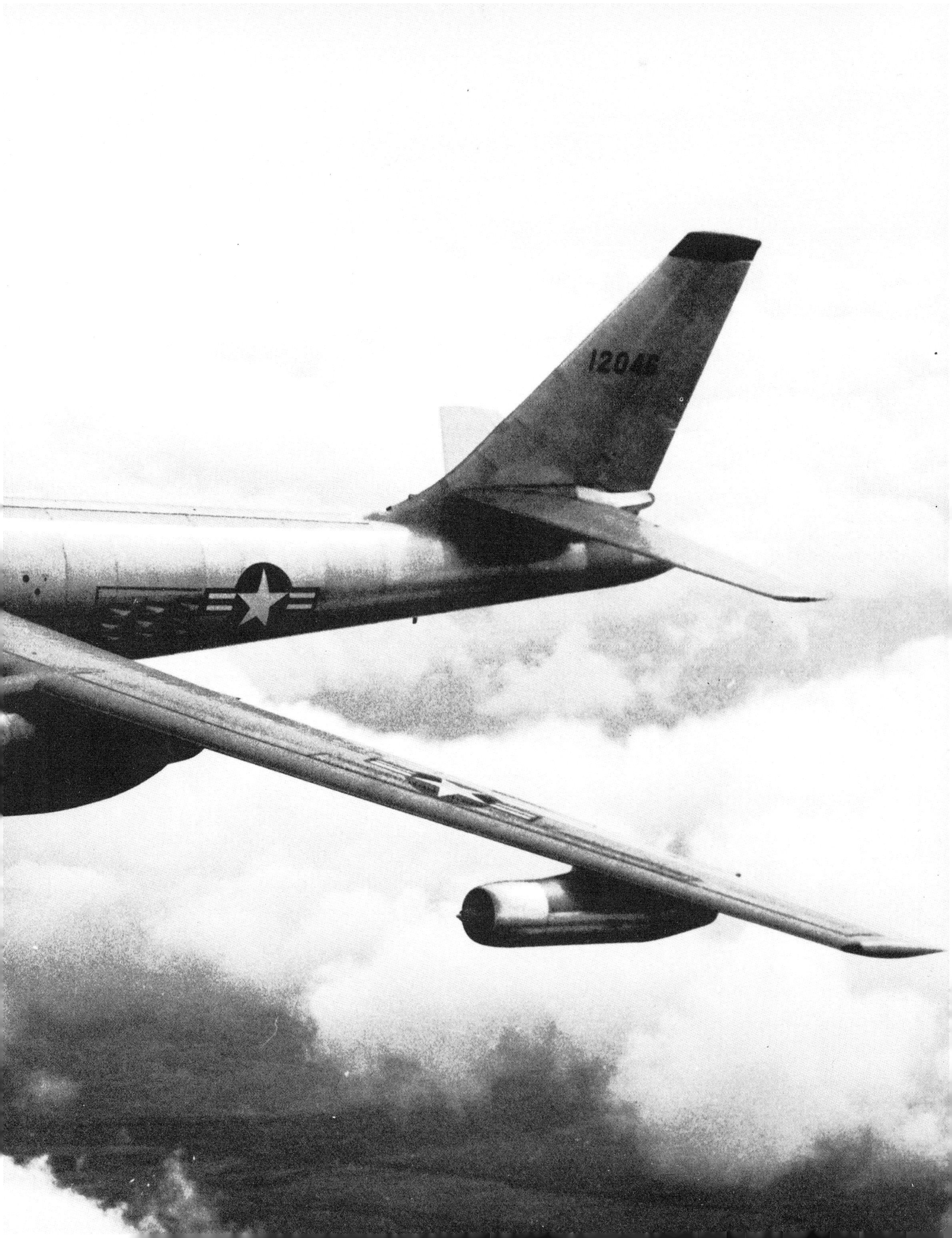
12046

Above: *The late, unlamented Rascal missile carried under a YDB-47E. Cynical programme officers on Rascal sometimes wondered if the missile would not occasionally fall up instead of down, just to be different.* (Boeing)

and failure due to stress corrosion of the "milk bottle pin", the main fitting holding the wing to the fuselage. There ensued a nightmare of fixes, new problems, further fixes and further problems. After an immense amount of effort in a close approximation of wartime urgency, Boeing, Lockheed and Douglas succeeded in an enormously complicated, immensely expensive programme that ultimately contained the problem.

The B-47s had another six years of service remaining, so the Milk Bottle Programme as it became known was worthwhile on its own merit. Even more important, it pointed out previously unconsidered problems inherent in the use of what were then considered exotic high strength metals for jet aircraft construction. There had been very little information available on possible fatigue problems with these metals, and the unknown ground being charted proved to be fraught with hazard.

On the positive side, the B-47 fatigue and stress corrosion problems led to improved maintenance and inspection methods which would enable the B-52 to endure even more severe challenges.

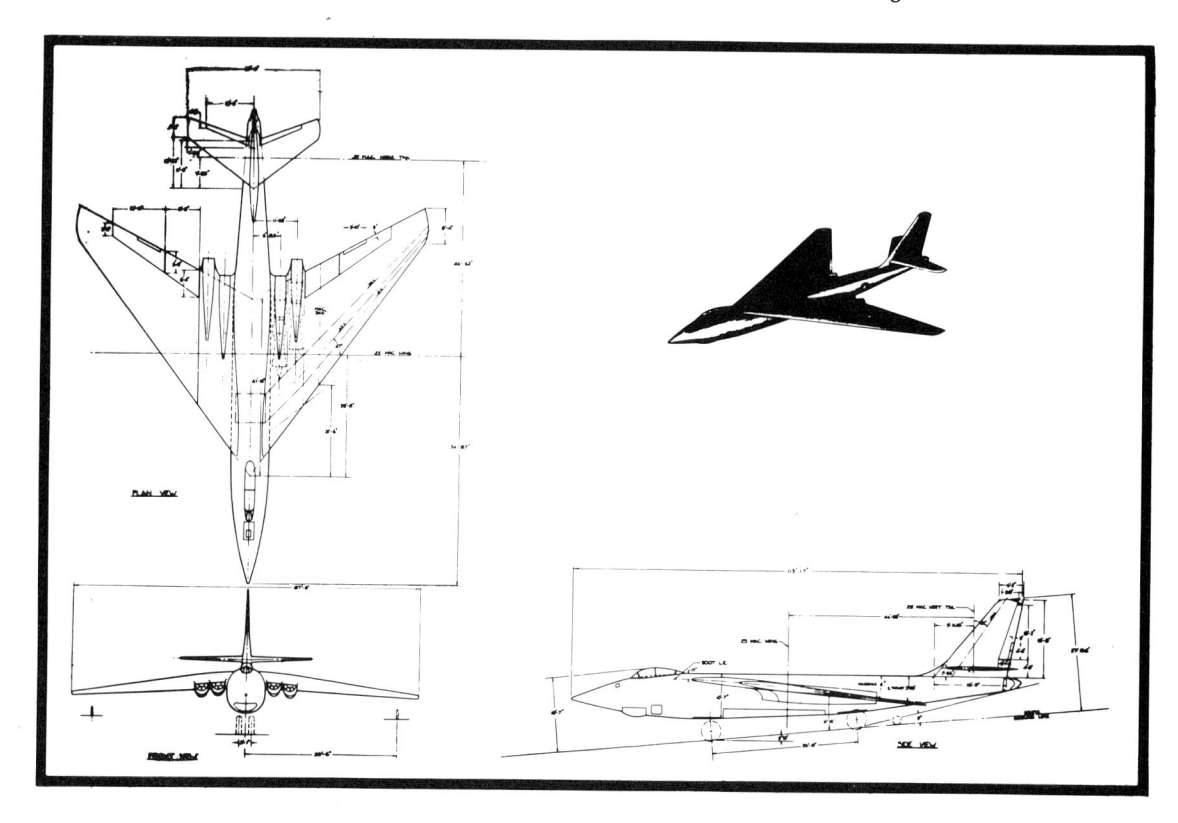

Left: *This Model 450-65-10C was one of several proposals for a supersonic follow on to the B-47. Engines were to have been J57s. Span was 87 feet, with 3.5 to 1 aspect ratio, and 2,190 sq ft area.* (Boeing)

THE DEVELOPMENT OF THE B-52

There were a whole series of events and circumstances which permitted the B-52 to be developed into the long-lived weapon system that it turned out to be. Some of it had to do big bomber experience and the general nature of the Boeing Company, whose managers and engineers had an enterprising, courageous Pacific Northwest frontier attitude that still permeates the firm with a palpable presence. Some had to do with the corresponding situation in the Air Force, where the bureaucracy had not yet developed to the point where it swallowed leadership, and where relatively junior officers were able to make key decisions which were fundamental to the B-52 development process. Similarly, senior officers were still permitted to exercise the vision, imagination and leadership which were then and are still the primary reasons for their existence.

There were other factors, too. Congress had not yet intruded so pervasively into the innermost details of aircraft procurement, and the public was able to countenance without apologies the purchase of a weapon system which promised a clear armed superiority over potential enemies.

Sadly, almost none of these conditions exist today; it is a case of too many people becoming "expert" in the process of aircraft design, and too many "safeguards" being built into the procurement process. These safeguards are of excellent intent, but of crippling effect. There are other problems, also. The mechanism of developing weapon systems has spawned innumerable subsidiary offices, many of which tend to become advocacy hobby shops, where well intended but career oriented personnel, civil and military, lose sight of the ultimate objective of the weapon system in their desire to see their own ideas adopted. Thus it was that basically excellent designs like the Lockheed C-5A and General Dynamics F-111 were saddled with concepts that were not relevant to the essential mission, resulting in cost growth and delays which harmed the final product and gave the programmes odious reputations.

The Congress, in its desire to know everything about a weapon system in real time, has put itself in the position of a restaurant customer checking in with the chef every step of the way, sampling, tasting, directing, changing his mind, and making decisions long before the menu is defined.

In simple terms, the conditions of the time of the B-52 development were such that the Boeing Company was given stringent requirements to produce a bomber with great range, speed and load carrying ability. When Boeing responded with a design that met these requirements and still had enormous growth potential, it was possible to avoid the operations research analysis which has cursed bomber development since, and which has "optimized" designs to the point where they are either incapable of growth (like the Convair B-58) or are so attenuated in their development programme that they become obsolete (or worse, out of fashion) before reaching quantity production, as in the case of the XB-70 and B-1.

The irony, of course, is that the billions pumped into the development of these aborted systems are then matched by billions pumped into desperation make-shifts like the FB-111. More money is spent, and no effective bomber with a growth potential is procured. Sad.

The Warp and the Woof

Each of the many threads which led to the quantity production of the B-52 is fascinating and they intertwine in a way which provides immense satisfaction today to the original civil and military participants in the process, which can be traced, idea by idea, back to the chill grey waters of the Atlantic in August 1941, in Placentia Bay, Newfoundland and to the first of the nine epochal conferences of Roosevelt and Churchill.

The Atlantic Charter which resulted from the meeting was a stirring statement of aims which covered joint global needs. One of the unreported decisions, however, was the mutual agreement that if Britain were invaded, and forced to continue the war from its Empire outposts, it would be necessary for the United States to build a bomber with a sizeable bomb load, capable of operating from the continental US against Germany.

The Army Air Force developed a requirement for what was really the next step in the programme which had already called forth the Boeing XB-15 and Douglas XB-19. The new aircraft was to have a 10,000 pound bomb load and a range of 10,000 miles. It had to be capable of carrying 72,000 pounds of bombs over shorter distances, have an airspeed at its 35,000 foot operating altitude of 250 to 300 mph and be able to operate from 5,000 foot runways. Formidable as this seemed in 1941, Boeing, Consolidated, Douglas and Northrop all submitted entries. Boeing was actually too preoccupied with B-17 and B-29 production to participate, while Douglas, based on its B-19 experience, came to feel that the

43

design was not attainable within the limits of current experience and dropped out.

The Consolidated design was selected, but there was also interest in the radical Northrop proposal, which ultimately became the XB-35 flying wing.

The development of the Consolidated XB-36, a 230 foot wing span, 276,000 pound gross weight aircraft took longer than the world events which removed the requirement for its existence. It first flew on August 8, 1946, long after both Japan and Germany had surrendered, and somewhat before the Cold War had spawned another need.

Despite tremendous inter-Service argument on the issue, in which neither the Navy nor the new Air Force appeared at their respective bests, the B-36 came into operational use with the Strategic Air Command in 1948, and built to a peak strength of 238 B/RB-36s by 1953. It was never to drop a bomb in anger, but its existence would on no less than seven occasions almost strangle the B-52 programme at birth.

Boeing had ended the war with numerous cancellations for its B-29 contracts and with only token orders for its successor, the B-50. It had a host of development programmes in parallel, however, the most important being the XB-47 and the next most a response to the Air Force's early 1946 requirement for a "second generation" successor to the B-36.

Turboprops

The requirements for the second generation aircraft were strikingly similar to the original set except that speed was to be 450 mph and the take-off distance, over a 50 foot obstacle, was lengthened to 7,500 feet.

Work on the XB-47 was well underway, but Boeing had not yet learned how well they had done with drag reduction on the six jet medium bomber. Conventional wisdom turned them to the design of the Model 462, a 221 foot wing span, straight wing aircraft powered by six Wright T35 turboprop engines of 8,900 shaft horsepower, with a gross weight of 360,000 pounds.

In the next two years, Boeing considered no less than 30 different combinations of engine, wing and gross weight to achieve the necessary speed and range requirements. None of the combinations worked, and worse, they failed against a background of the steadily increasing capability of the B-36, which, in its D model, with six 3,500 horsepower Pratt & Whitney R-4360 radial engines and four General Electric J47s suspended in B-47 pods, seemed to almost meet the Air Force requirement. None of Boeing's second generation aircraft exceeded the B-36's performance in any significant way, nor did they offer much in the way of growth potential.

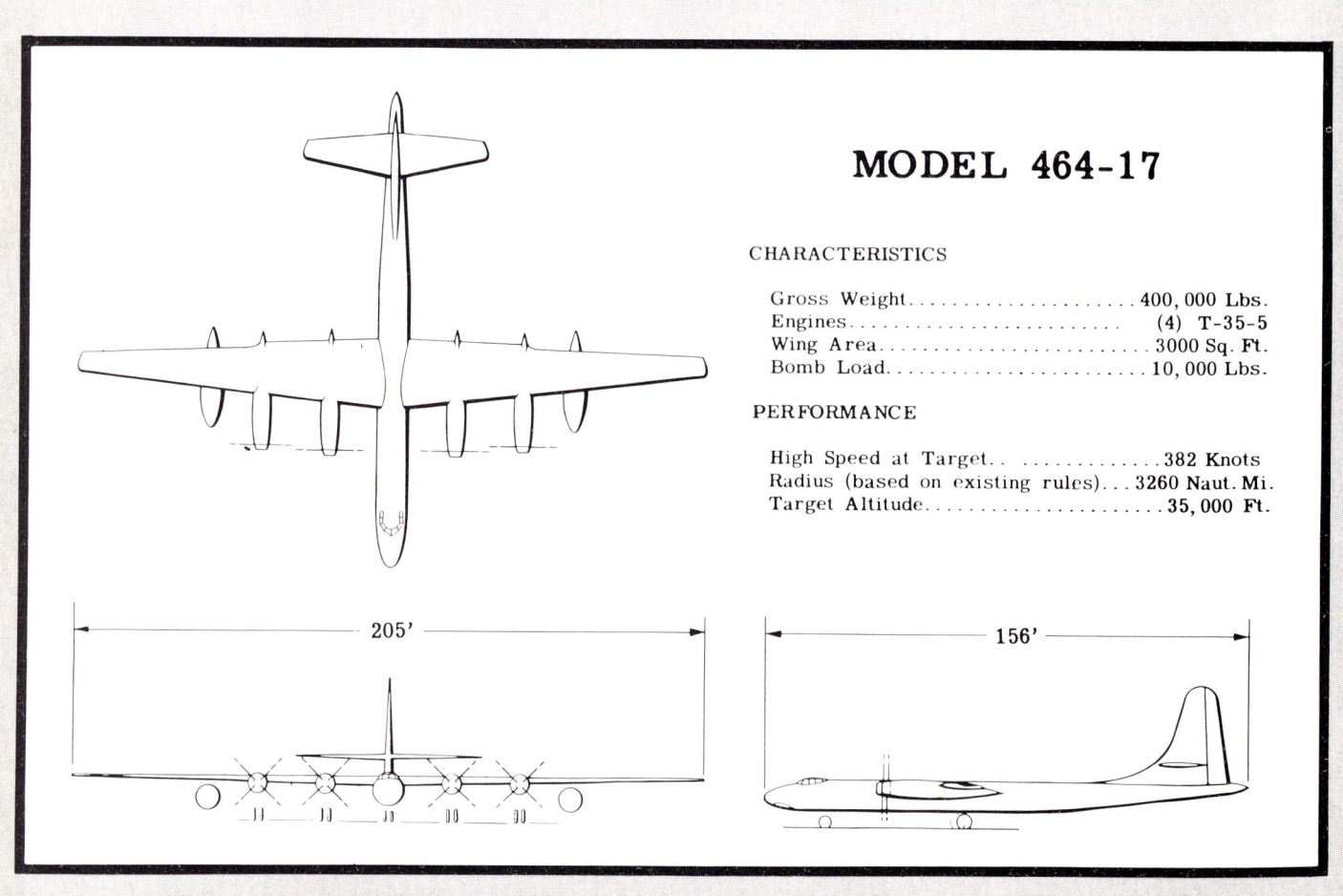

Pages 44-7: The progression of the B-52 design through a few of its many configurations, showing the changes in size, weight, and powerplant configurations. (Boeing)

MODEL 464-17

CHARACTERISTICS

　　Gross Weight.....................400,000 Lbs.
　　Engines.........................(4) T-35-5
　　Wing Area......................3000 Sq. Ft.
　　Bomb Load......................10,000 Lbs.

PERFORMANCE

　　High Speed at Target...............382 Knots
　　Radius (based on existing rules)...3260 Naut. Mi.
　　Target Altitude....................35,000 Ft.

205'

156'

MODEL 464-35

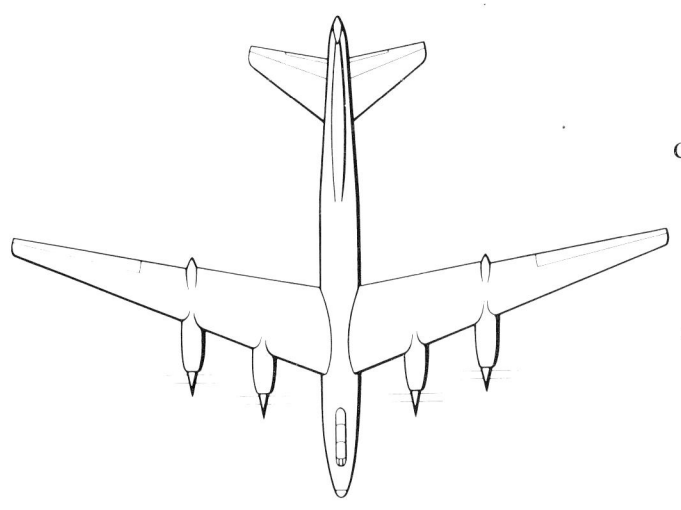

CHARACTERISTICS

Gross Weight.....................280,000 Lbs.
Engines........................ (4) T-35-W-3
Wing Area...................... 2600 Sq. Ft.
Bomb Load (N.M.E.).............10,000 Lbs.

PERFORMANCE

High Speed at Target................435 Knots
Radius (based on N.M.E. rules)...3070 Naut. Mi.
Target Altitude......................41,000 Ft.

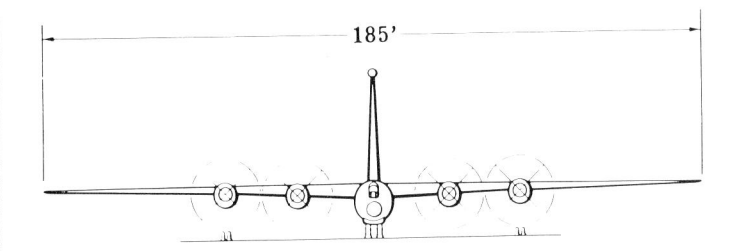

185'

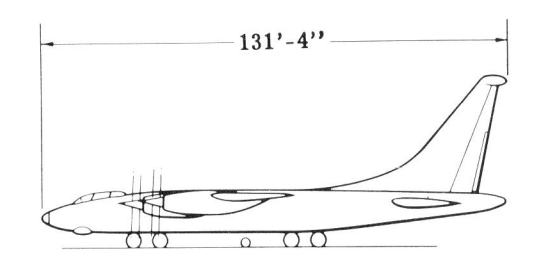

131'-4"

MODEL 464-40

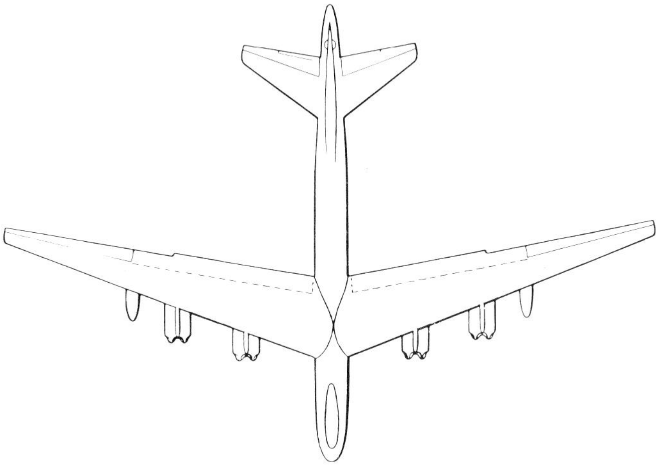

CHARACTERISTICS

Gross Weight.....................280,000 Lbs.
Engines.................... (8) XJ-40-WE-12
Wing Area...................... 2600 Sq. Ft.
Bomb Load (N.M.E.).............10,000 Lbs.

PERFORMANCE

High Speed at Target................440 Knots
Radius (based on N.M.E. rules)...2660 Naut. Mi
Target Altitude.....................47,400 Ft

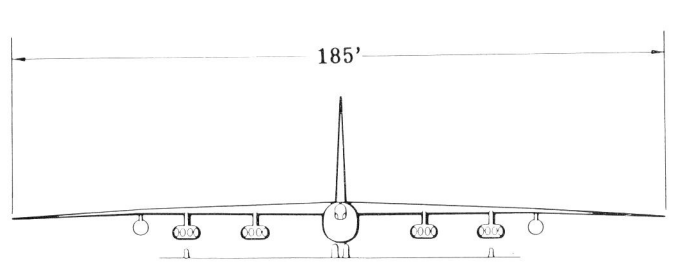

185'

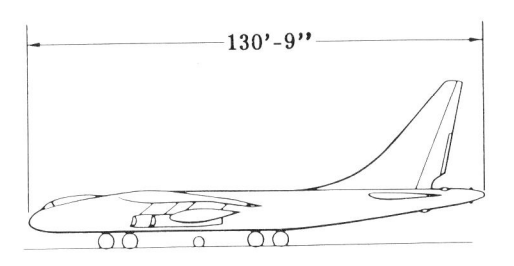

130'-9"

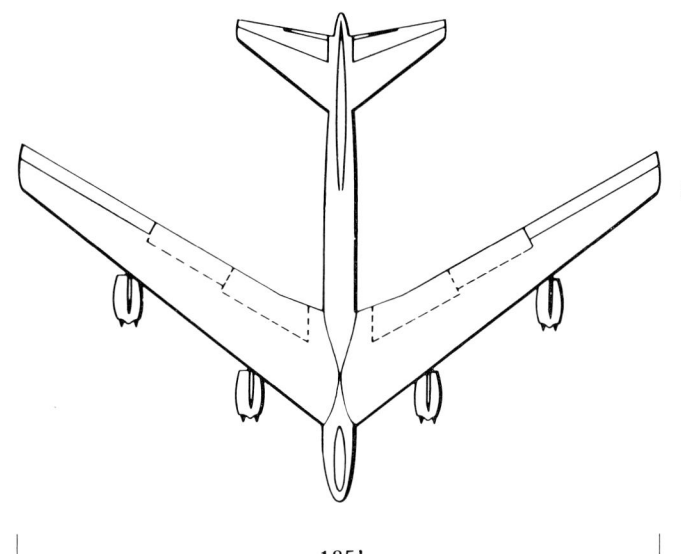

MODEL 464-49

CHARACTERISTICS

- Gross Weight.....................330,000 Lbs.
- Engines.........................(8) J-57
- Wing Area.......................4000 Sq. Ft.
- Bomb Load (N.M.E.).............10,000 Lbs.

PERFORMANCE

- High Speed at Target...............490 Knots
- Radius (based on N.M.E. rules)...2660 Naut. Mi.
- Target Altitude....................49,400 Ft.

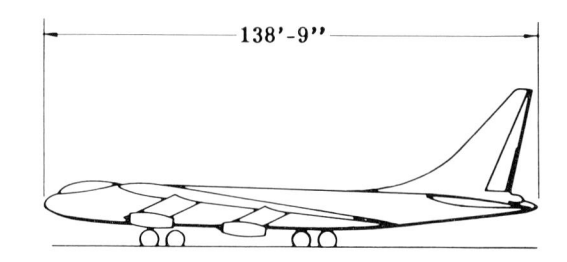

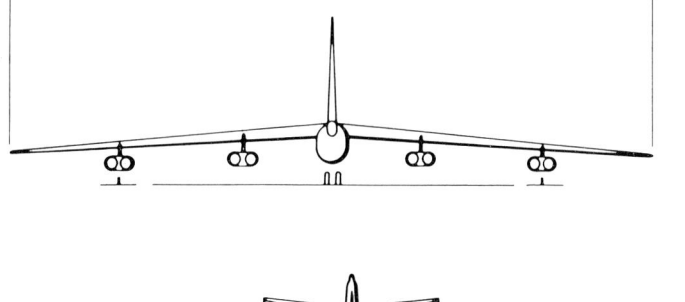

185' 138'-9"

MODEL 464-67

CHARACTERISTICS

- Gross Weight.....................390,000 Lbs.
- Engines.........................(8) J-57 P
- Wing Area.......................4000 Sq. Ft.
- Bomb Load (N.M.E.).............10,000 Lbs.

PERFORMANCE

- High Speed at Target...............491 Knots
- Radius (based on N.M.E. rules)...3070 Naut. Mi.
- Target Altitude....................46,500 Ft.

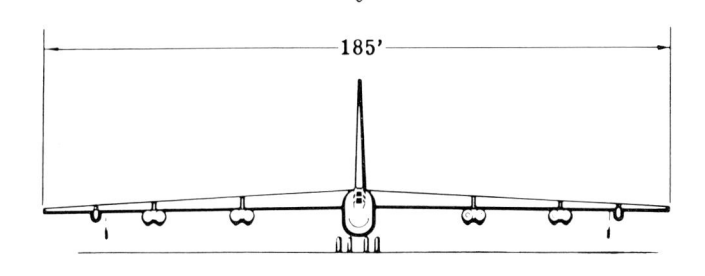

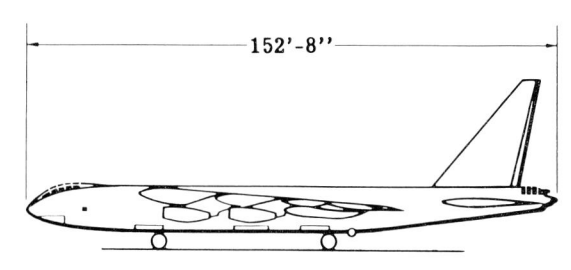

185' 152'-8"

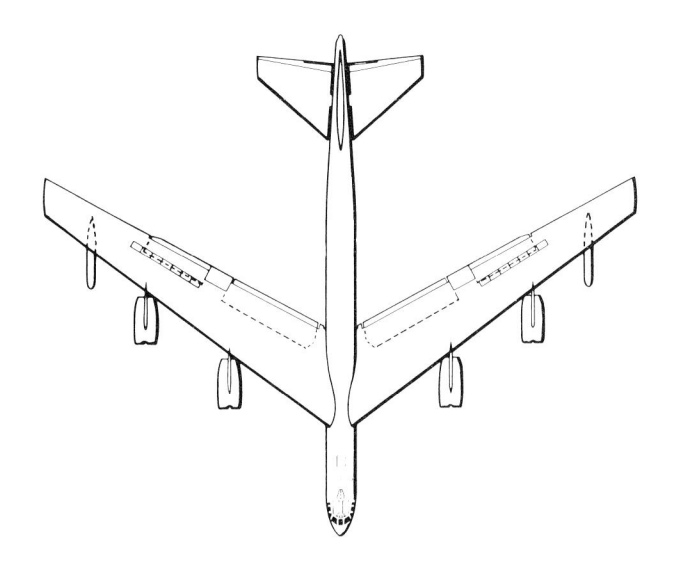

MODEL 464-201

CHARACTERISTICS

Gross Weight	390,000 Lbs.
Engines (8)	J-57-P-1
Wing Area	4000 Sq. Ft.
Bomb Load (N.M.E.)	10,000 Lbs.

PERFORMANCE

High Speed at Target	490 Knots
Radius (based on N.M.E. rules)	3160 Naut. Mi.
Target Altitude	46,700 Ft.

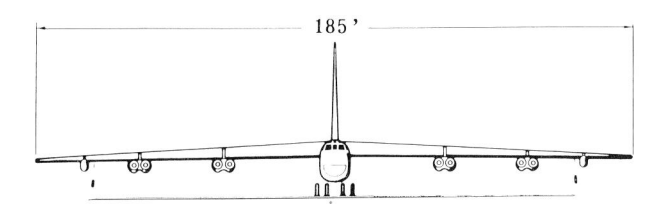

185'

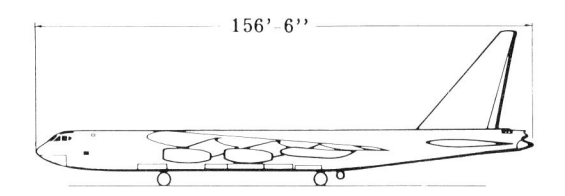

156'-6"

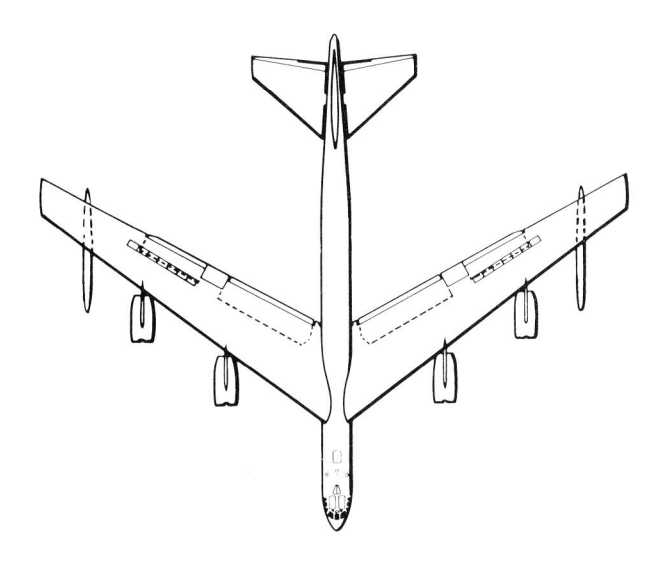

MODEL 464-201-6,-7

CHARACTERISTICS

Gross Weight	450,000 Lbs.
Engines (8)	J57-P-19W
Wing Area	4000 Sq. Ft.
Bomb Load	10,000 Lbs.

PERFORMANCE

High Speed at Target	496 Knots
Radius (Based on MIL-C-5011A Rules)	3325 Naut. Mi.
Target Altitude	45,050 Ft.

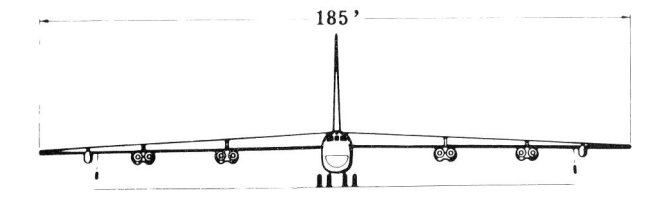

185'

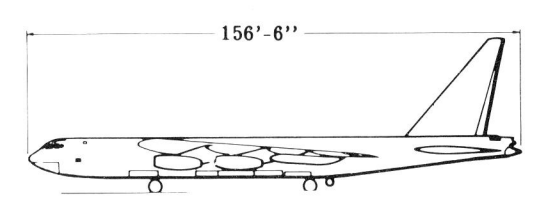

156'-6"

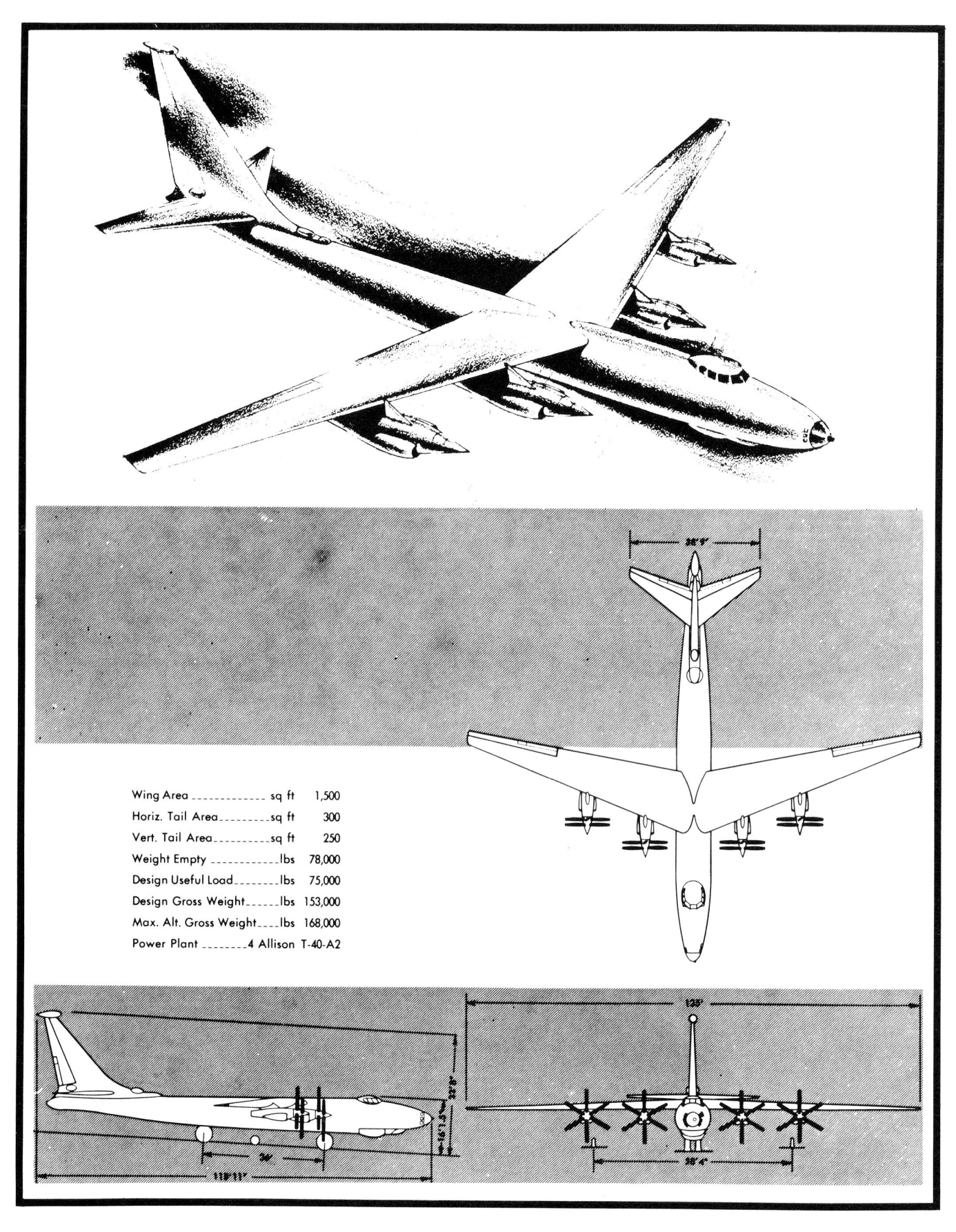

Wing Area _____ sq ft   1,500
Horiz. Tail Area _____ sq ft   300
Vert. Tail Area _____ sq ft   250
Weight Empty _____ lbs   78,000
Design Useful Load _____ lbs   75,000
Design Gross Weight _____ lbs 153,000
Max. Alt. Gross Weight ____ lbs 168,000
Power Plant _____ 4 Allison T-40-A2

Left and above: *There were simultaneously being developed at Boeing numerous other projects, including medium bombers like this turboprop Model 474, with its aft located gunner's compartment.* (Boeing)

Below: *The Model 474 matured into the Model 479, powered by six J-40 jet engines and featuring a thickened wing root section that would reappear in the B-52. This became the competition winning XB-55, the intended successor to the B-47. With the almost simultaneous development of the B-52, the XB-55 was shelved.* (Boeing)

The Necessary Conditions

In a single October weekend in Dayton, Ohio, a whole series of separate events, each of which had followed a complex development path of its own, were focused together like rays of the sun by a small team of Boeing and Air Force personnel.

These were the separate events, some with very long histories, which came together to give birth to the XB-52:

(1) The theory of inflight refuelling had become a fact, and there were major improvements (the Boeing developed flying boom) which promised to increase vastly the efficiency of the process.

(2) The XB-47 drag had proved to be much lower than anticipated, and much had been learned about the aerodynamics of high speed flight.

(3) Such was the enthusiasm for the XB-47 that Boeing had on July 1, 1948 already been granted contract W 33-039 D-21777, calling for its successor, the XB-55. This medium bomber, numbered MX-1022, had already gone through almost as many concept changes as the XB-52, evolving from a straight wing four turboprop effort to what later looked very close to a B-47 with four turboprops instead of jets, and into a much modified four jet version of the B-47. (A supersonic delta wing aircraft design was also an offshoot of this programme.)

(4) The company developing the turboprop engine could not agree with the company developing its propeller as to whether the shaft of the engine was strong enough to sustain the forces induced by the propeller at high RPM, but was able to agree that it would take at least four years before there would be a successful powerplant.

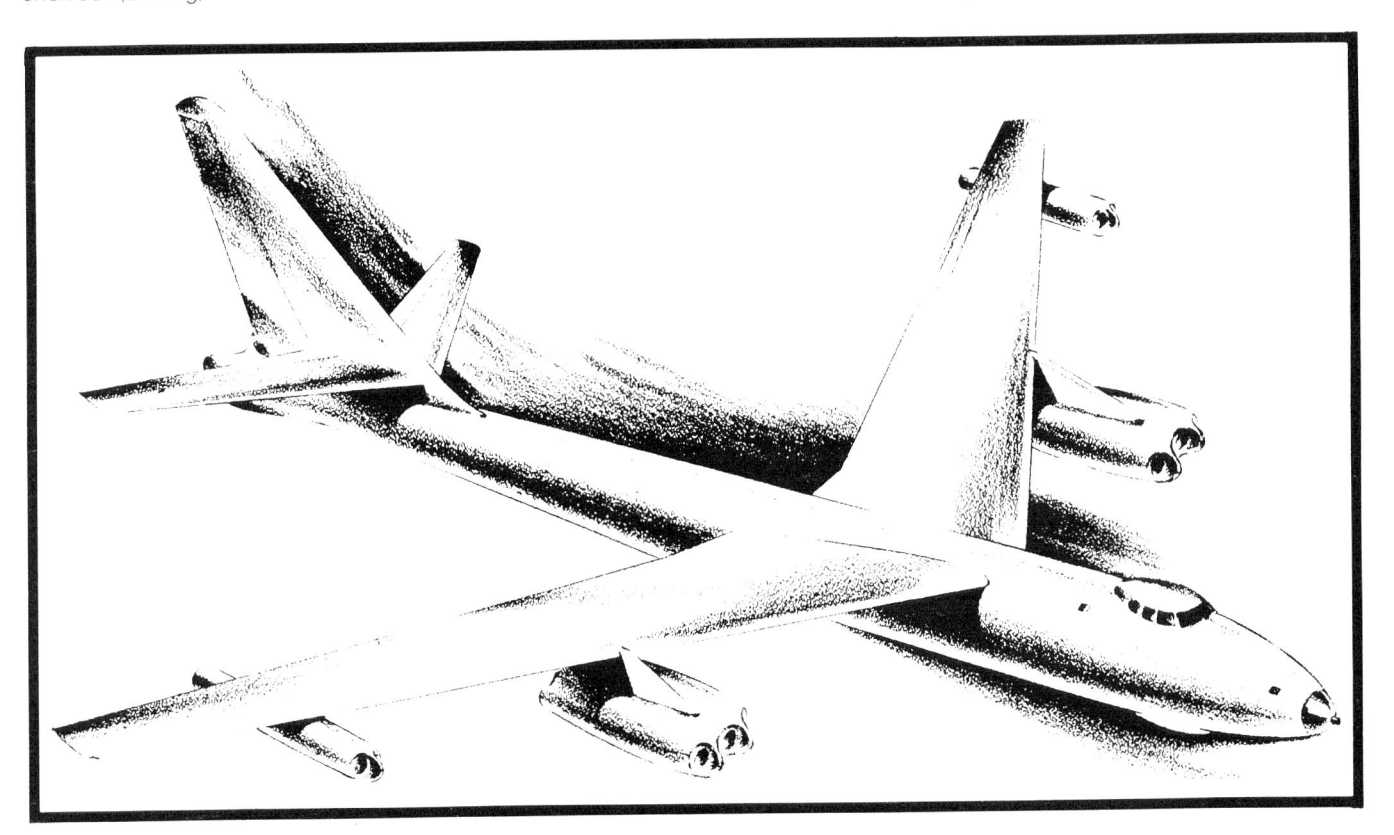

(5) In the process of developing the B-47 there had grown in the Air Force a keen awareness at the highest levels that a swept wing jet bomber was clearly the way of the future. From the Commander of the Strategic Air Command, Lieutenant General Curtis E. LeMay, to the Commander of the Air Materiel Command, Major General K. B. Wolfe, to the Chief of Bomber Development, Colonel Henry E. "Pete" Warden, there was a firm understanding that the B-47, for all its promise, was simply not big enough to accomplish all that might eventually be required of it. (At other levels in the structure, however, there was much opposition to this philosophy.)

(6) There was within the Boeing organization sufficient flexibility to have an interchange of ideas from all of the many bomber proposals, and sufficient imagination to put them together in a winning combination.

First Prize: A Weekend in Dayton

All of these separate strands of development came together during the week of October 22, 1948, in Dayton, Ohio. Colonel Pete Warden had, somewhat outside the ordinary bounds of his authority, been urging Pratt & Whitney to develop what became the J57 engine.

On Thursday of that fateful week, during lunch, after a long discouraging morning filled with nothing but bad news about the prospects for the turboprop engine, Warden met with senior Boeing personnel on the problem. Warden suggested that they scrub the turboprop bomber entirely, and look into the prospects for a swept wing pure jet aircraft.

The Boeing engineers went back to the Van Cleve hotel and in classic "back of the envelope" style began synthesizing the years of effort that had accumulated on all of their diverse programmes into one. On Friday morning they called Warden and told him they would have a proposal on the following Monday.

It happened that the very best minds from Boeing were in Dayton that weekend, and they alerted their colleagues in Seattle to stand by to provide data by phone as needed. The group included Edward C. Wells, George S. Schairer, H. W. "Bob" Withington, Vaughn Blumenthal, Art Carlsen and Maynard Pennell, all of whom would go on to distinguished careers with the company.

Below: *Some of Ed Well's pencil sketches during the days preceding the B-52 design. They are remarkable both for how close they came to the final configuration, and how radical were some of the departures.* (Boeing)

Right: *These are copies of the original XB-52 drawings submitted in the famous "hotel room" proposal. The effort represented a distillation of current Boeing thinking.* (Boeing)

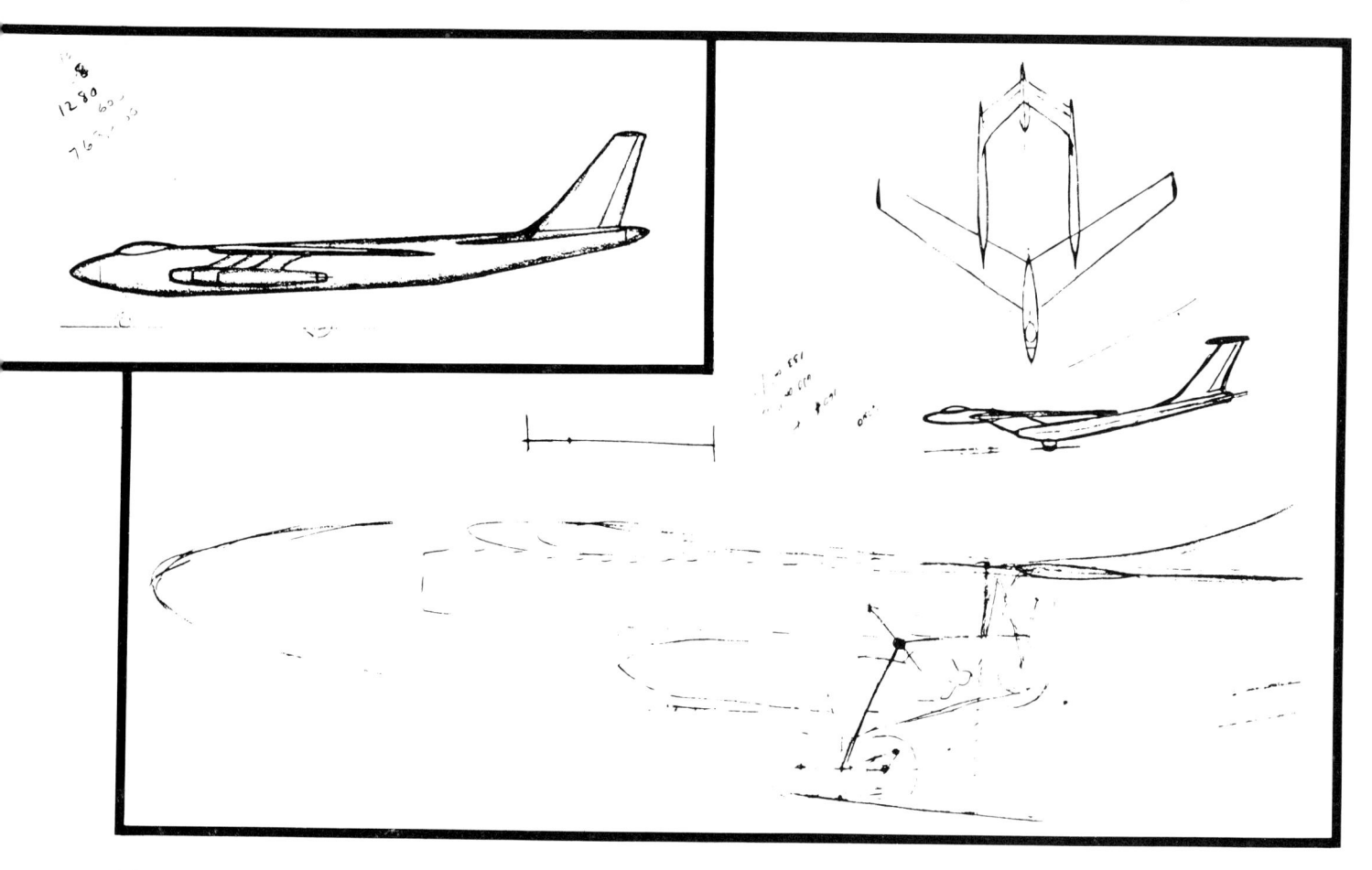

These six men distilled their wisdom into an entirely new aircraft design, the original concept of which was strikingly similar to the huge aircraft which rolled out of the Boeing plant for ground tests on November 29, 1951. They created a 33 page proposal for a large aircraft, with 4,000 square feet of wing area, 35 degrees of wing sweep, eight Pratt & Whitney J57 engines in B-47-like pods, and a slim low drag angular fuselage with the pilots seated in tandem. Called the Boeing Model 464-49-0, it had a design gross weight of 330,000 pounds, a high speed of 572 mph and a range of 8,000 miles with a 10,000 pound bomb load.

On Monday morning the team presented Warden with the slim proposal (which in today's environment would have to constitute a freight car load of data) which included an inboard profile, three-view drawing, drag polars and weight estimates.

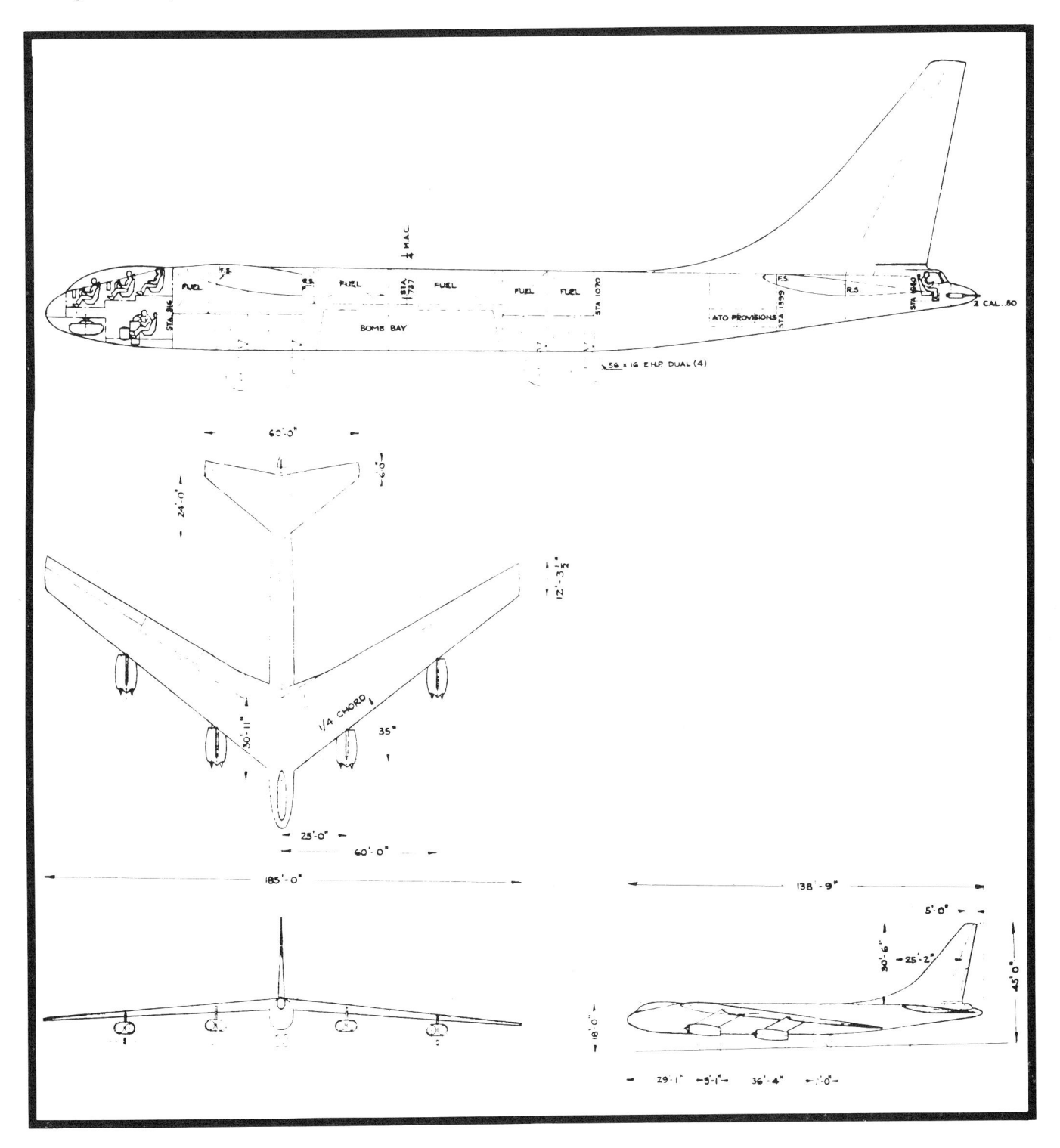

Ed Wells, already a world famous engineer with extraordinary status in the aviation community, not only did the three-view drawings, but assisted Schairer with the construction of a balsa model of the proposed aircraft, which, painted silver, was put on a stand and presented to Warden to take back to the Pentagon.

Warden was ecstatic with the proposal despite the fact that it combined a new airframe with new engines and a new technique of inflight refuelling. Acting on his own authority, confident that he would receive backing from his superiors, he authorized Boeing to terminate their efforts on the turbo-prop projects, and promised to deliver new funding for the XB-52 within a few months. And he did.

Similarly, Pratt & Whitney was induced to go ahead with their J57 development, a development which also made possible a generation of civil jet transports.

Below: *The unlikely competitor, Convair's YB-60. Essentially a swept wing jet powered version of the B-36, it didn't worry Boeing technically, but there were political ramifications that were always troublesome.* (Convair)

Impact

The crucial importance of the new bomber was immediately recognized by General LeMay, who wanted it in production at the earliest time, and who had made up his mind to stop further development of the B-47 because he did not want it to compete for funds.

LeMay stories abound at Boeing, but George Schairer, Chief Aerodynamicist at the time, tells one about LeMay putting his arm around him at a meeting and saying "George, whatever you are doing to improve the B-47, stop it". Guy Townsend, who was test pilot on both the XB-47 and XB-52 programmes, tells a saltier one. A team from Wright Field showed up at Omaha to brief LeMay on a proposed new version of the B-47. Powered by four J57 engines and with no other changes, it would equal the XB-52 range. LeMay reportedly stopped the luckless briefing officer before he started with a single question "Just how deep does a programme have to be buried before you dumb sons-a-bitches at Wright Field will stop digging it up?"

Yet as attractive as the XB-52 seemed to be to the principals in the programme, it ran into internal Air Force resistance at several levels. The powerplant, armament and propeller divisions at Wright Field all opposed it because it had leap-frogged their current development programmes. It ran into opposition from people managing the facilities funds because it would occasion some mammoth construction efforts that had not previously been budgeted. It in fact ran into the same opposition from entrenched parties that new ideas always do.

Maynard Pennell and Art Carlsen had put together the weight statement for the proposal, and because Carlsen was project manager he made every effort to adhere to it. The real production problems with the B-52 would not come with the airframe itself, but from the innovative systems that were adopted to reduce weight. Where the B-47 had in almost every instance used proven off-the-shelf systems, the B-52 was virtually a hot house of new ideas. It was this conscientious approach to reduce drag and empty weight that was a major reason for its ability to serve so well, so long.

As simple as it sounds, the combination of low empty weight, low drag, large wing area (with the resultant high L/D [Lift over Drag] ratio) and large size of the airframe was the key to the aircraft's success.

Even so, the Air Force saw a need to further increase the range of the aircraft. In October 1949, Warden requested Blumenthal and John Alexander to come to Dayton, Ohio to explore the possibilities. An in-depth review was made to each of the Wright Field Laboratories during the month of October and resulted in the aircraft's weight being increased to 390,000 pounds officially. (And 405,000 pounds unofficially—Warden liked to keep a 15,000 pound weight "kitty" in his back pocket.)

Boeing, in its second attempt at a production jet bomber, did so well with the available knowledge that there has been no subsequent subsonic bomber which could replace it. It has outlived the B-58, the B-70 and the B-1, and with the new systems still being installed is a strong candidate to outlive the FB-111 and any B-1 successor.

The author asked Ed Wells, George Schairer, Vaughn Blumenthal and George Martin what they might have done differently with the B-52 if they had had an opportunity now to change it. Each of the men thought the question over carefully, and each came to the conclusion that given the knowledge available at the time, they would not change the design in any way.

Characteristics of Early Aircraft

Once the key configuration had been reached in Dayton, events came with almost bewildering speed. A comprehensive 670 day wind tunnel test programme was complemented by a similarly intensive theoretical testing of the design. Several crucial elements emerged, some of which were contrary to contemporary Air Force thinking.

The Wing

The XB-47 had been a relatively low budget programme, and its manager, George Martin, says that initial calculations had led them to believe that the thin wing was critical at high Mach numbers, and that the wing/fuselage juncture was an especially important area. As a consequence, the structure of the very thin wing had to be exceptionally heavy, and there was not much room for fuel storage.

Further wind tunnel studies revealed, however, that the most critical place for drag rise was somewhat further out on the wing, and this was translated first in the XB-55 on paper and then in the XB-52 in practice as a much thicker wing root section. This had two immediate advantages plus a third that was discovered later. First, it permitted a much lighter construction with a consequent reduction in overall weight. Second, it permitted the storage of fuel in the wing, which had the effect of what is now called "span loading" and again permitted a structural weight reduction. The third item, whose value was only apparent later, was that it facilitated the design of the modern Boeing jet commercial transport. While a bicycle type landing gear was acceptable in a bomber, which required less space for a bomb bay than a transport did for passengers, it rendered the design of the transport impractical because so little room was left for seating. The thickened wing root, on a low wing aircraft, would provide an area to store a conventionally located landing gear and provided the way for the 707 configuration.

Vaughn Blumenthal, who became a leading aerodynamicist at Boeing, and John Alexander were called to Wright Field to justify this thick wing approach, which was counter to Air Force thinking. Colonel Warden hauled them in to talk to Dr Goethart, a German scientist who had come over in "Operation Paperclip". Dr Goethart listened to their presentation, then rummaged around among some old papers written in German, made a series of calculations, and then announced that yes, it was theoretically possible if the point of maximum thickness was swept as you went out in span. Outside confirmation convinced the Air Force, and Boeing was given a go-ahead.

The huge 185 foot span wing was actually quite thin, although appearances were deceiving. On the centre line of the fuselage, the wing structure had a thickness ratio of 16·2%. At 25% of span this dropped to a 10·3% figure, which further declined to 9·4% at 57% of the span, and to only 8·0% at the tip. The aerofoil was modified from a NACA 64 series at the root to a 66 series at the tip.

The thinness of the wing resulted in a flexibility similar to the B-47's; the maximum wing tip up to wing tip down deflection was 32 feet, seemingly impossible for such a massive structure.

Boeing had felt that perhaps the one basic mistake it had made with the B-47 was giving it only 1,400 square feet of wing area. The B-52 had 4,000 square feet, and like cubic inches in racing car engines, wing area is the *sine qua non* of long range performance. The range equation was further helped by the high 8·55 to 1 aspect ratio.

The pod mounted nacelles that had worked so well on the B-47 made the same sort of contributions to the B-52. They were placed so as not to add to the drag rise at high Mach numbers, served as load alleviation, and helped mitigate the stall. The nacelle struts served as fences and were tuned to avoid the possibility of flutter.

Lateral control on early models of the aircraft was effected by means of spoilers and ailerons. The aerodynamically balanced ailerons were small and manually controlled by the pilot by means of a control tab. The ailerons supplied most of the rolling power and were adequate by themselves for normal operation. For landing, refuelling or "combat type"

Above: *One of George Schairer's biggest contributions to Boeing was his insistence on the building of a very modern wind tunnel. It was to prove invaluable in the B-47, B-52 and numerous other programmes. Note socked shoes and the flexibility of wings.* (Boeing)

manoeuvres, additional control is supplied by hydraulically operated spoilers, which deflect in proportion to control wheel movement. The spoilers had the great additional advantage of acting as drag brakes when applied symmetrically, thus eliminating the need for an approach chute. A 44 foot diameter brake chute was still required for stopping.

Fuselage

While not an aesthetic triumph, the B-52 fuselage was a spacious low drag structure with very little wetted area for its size. The XB and YB-52s were fitted with tandem seating like the B-47s, for drag reduction purposes. LeMay insisted for reasons of crew coordination that the seating be made side by side in production aircraft, and approval for this was given in August 1951.

Despite the aircraft's apparent size, the crew compartment is much smaller than would be imagined, and is crammed with equipment. In all production models the aircraft commander and pilot sit side by side, while the Electronic Counter Measures Officer (or Electronic Warfare Officer, as he was called later) is situated in the aft portion of the cabin upper deck. On the lower deck, the radar navigator and navigator share associated equipment. The tail gunner was housed in a separate pressurized compartment in all models down to the G, when he was relocated forward next to the Electronic Warfare Officer.

The bomb bay is 28 feet long and six feet wide, and at its introduction and since could accommodate any weapon in the US arsenal. On certain aircraft, RB-52Bs and B-52Cs, a special pressurized capsule could be inserted in the bomb bay to provide additional reconnaissance capability, either electronic or photographic.

Empennage

As tall as the 48 foot high tail of the B-52 appeared, it was smaller than had been planned for earlier proposals. The surface could be folded for maintenance, and the rudder had a very narrow chord. Later in the G and H models, the height would be reduced to 40 feet.

The most important element of the empennage was the trimmable stabilizer, the first ever used on an aircraft of its

Above: *Despite its enormous size and strength, the B-52 wing was even more flexible than the B-47s, moving in an incredible 32 foot arc during static tests. The aircraft had a decided anhedral while on the ground, and assumed a different degree of dihedral for every weight during flight.* (Boeing)

size, and a vital improvement over the B-47. The stabilizer operated through 13 degrees of motion by a hydraulically driven irreversible jack-screw. Stabilizer trim checks were of crucial importance before take off, for the relatively small elevators (only 10% of stabilizer chord) could not overpower the stabilizer. At least one aircraft was lost in a take-off accident because of improperly set trim.

The trimmable stabilizer made it possible to rotate the aircraft on take off, rather than having to fly it off the ground at a predetermined attitude as was the case with the B-47. It also permitted flare in the landing approach to touchdown.

Auxiliary Systems

Pneumatic The use of a pneumatic system on the early models for the primary power source in the operation of all auxiliary functions was a dramatic departure from convention, and its benefits in lighter weight were only partly offset by the need to carefully install and maintain the ducts which took the extremely hot air (as much as 500 degrees F) past fuel cells, hydraulic lines, and so on. High pressure, high temperature air was bled from the second stage compressor in each jet engine, and carried by the ducts to the desired locality in the aircraft where it was transformed into electrical or hydraulic energy by air turbine driven power packs.

Hydraulic The ten turbine driven hydraulic pumps supplied pressure at 3,000 pounds per square inch to drive suitably bled down pressure for brakes, steering mechanism, bomb bay doors, spoilers and the adjustable stabilizer. Each pack was self contained, incorporating electric-hydraulic starting, speed control, air pressure control, and output regulation. The multiple packs reduced weight and vulnerability.

Electrical The original electrical system consisted of four air turbine driven 60 KVA alternators furnishing 200/115 volt three phase 400 cycle alternating current to all the major electrical functions. DC voltage was obtained by means of transformer rectifiers. The air turbines proved troublesome and after a series of failures, one of which cost a B-52 at Castle Air Force Base, were replaced on later models with gear driven constant speed drives and hydraulic pumps.

De-icing Bleed air was also used originally for the thermal de-icing of the leading edges of the wings, horizontal stabilizer and vertical fin. Guy Townsend proved in a series of tests that it was almost impossible to accumulate structural ice on the B-52, because the constant flexing of its surfaces removed it as fast as it formed. De-icing was then eliminated on all but the nacelles and engine inlets.

Landing Gear The four strut, eight wheel landing gear of the B-52 was highly secret at first, and early photos of the aircraft had the undercarriage airbrushed out. The quadricycle main gear was relatively light weight and provided a very short turning radius. The tip protection gears (outriggers) which retracted into outer wing panels, provided against dragging a tip in a wing low landing, or high speed ground manoeuvring. The front main wheels were steerable, and an excellent crosswind steering mechanism had 20 degrees of turn to either side, permitting aircraft to be crabbed into the wind while the wheels remained aligned with the runway. In a strong crosswind, pilots took a little time to get used to viewing the runway through a side window instead of through the windscreen.

Armament The aircraft was designed to have four ·50 calibre machine-guns fitted in a radar controlled tail turret. The gunner in A through F model aircraft was in a separate pressurized compartment, which could be jettisoned for bailout. Crew morale and weight reduction considerations dictated his move forward in the B-52G and H aircraft where he operated his weapons by remote control, and escaped the punishing ride experienced in the tail of the aircraft during turbulent flight conditions. In the H model, a 20 mm six barrel "Gatling" gun was installed. A later chapter will cover more on gun and missile armament.

Engines Pratt & Whitney had been excluded from jet engine development during the war, the US government not wishing its attention to be diverted from the main task of delivering the finest air cooled piston engines in the world. Some limited development work had been undertaken as early as

1939 on gas turbine applications, and the original focus had been on a propeller driven turbine type known as the PT1, powered by gas generated in a free-piston type Diesel gas generator.

Although this first effort was not successful, a second project, the PT2, introduced a multi-stage axial flow compressor. The PT2 ultimately was used on the Douglas C-133 transport with the T34 designation.

In 1947, however, Pratt & Whitney offered to develop for the USAF an advanced rotor axial flow turbojet engine for the B-52. Air Force interest, however, was reserved for the turboprop, and the company was advised to develop a turboprop engine of the same size. A contract was let for the development of a 10,000 horsepower PT4 engine (T45). Especially designed for the B-52, the programme was placed under the direction of veteran P&W engine designer Andrew V. D. Willgoos.

The PT4 employed a dual axial flow compressor of 13 stages, with a compression ratio of 8 to 1, and was designed so that it could easily be converted to a pure turbojet should the need arise. Before the two prototype PT4 engines were completed, the Air Force revised its requirement, and design work began on the derivative JT3 (J57) engine which was completed and run in June 1949.

Engine tests called for revisions and the first jet engine in aviation history to generate 10,000 pounds of static thrust was the JT3A, which ran in January 1950.

The first installation of this engine in the B-52 proved to be a complete success, and the engine was adopted for many new aircraft. For his leading role in conceiving of and developing the J57, Leonard S. Hobbs, Vice President for Manufacturing for United Aircraft Corporation, was awarded the Collier Trophy in 1952.

A subsequent variation, the JT3D, became the engine selected for the Boeing 367-80 prototype jet transport, and set the stage for the jet age revolution of air transportation.

The development of the J57 series put Pratt & Whitney Aircraft in the forefront of jet engine manufacture, and characteristically, the jet engines were continually improved. A major improvement was the introduction of the turbofan to the J57 configuration, resulting in the TF33 engine. Used in the B-52H, the TF33 ushered in the age of fan jets (turbofans) which permitted an increase in both power and economy.

The Momentum Builds

The establishment of the initial B-52 configuration was but one step in the massive programme to get the Stratofortress (as it is almost never called) into production. The Air Force, contrary to today's "fly before buy" policy, ordered production quantities before the prototype flew.

On the one hand, the Air Force had to do the planning and procurement necessary to identify bases where the aircraft would be stationed, lengthen and strengthen runways and parking aprons, build the necessary hangars and support facilities, specify the spare part procurement plan, set up the necessary air and ground crew training programmes, integrate the new ground and test equipment into existing inventories, provided for the various levels of maintenance, print the necessary handbooks, and look into the future to see what modifications, new equipment and so on would be necessary.

On the other hand, Boeing had to set up the system of subcontractors who in some cases would have to invent, design and manufacture the totally new items of equipment the B-52 called for. It had to train workers, order machinery and materials, plan for a second procurement source, maintain a scrupulous drawing control system, and like the Air Force, look into the future to see what modifications would be required.

The programme called for new bombing and navigation systems. New simulators for crew training, far more complex than anything previously purchased, had to be designed and built. An enormous series of orders flowed to all parts of the country for materials, parts and labour. The machining required to build the massive new structures was enormous, and at several points in the B-52's production and modification programmes, a great percentage of available US machining capacity was dedicated to it.

In the last analysis, however, the success or failure of any aircraft is indicated when the wheels finally leave the ground. It would happen on April 15, 1952.

Below: *The tarpaulin shrouded roll out of the first B-52, with a typical rainy Seattle background. The fin is folded. This photo was so dramatic that Boeing photographers later had the apron wetted artificially on some occasions, to catch the reflected light and heighten the drama of a night shot. (Boeing)*

The degree of grin depends a lot on who is visiting. In the first picture, General Curtis E. LeMay, Commander of the Strategic Air Command, tends to keep everyone serious. LeMay, who had the vision to demand an aircraft of the B-52's size, flew the aircraft very well from the Aircraft Commander's seat on his 85 minute first flight. From left, Richard Loesch, Boeing test pilot, LeMay; Lt. Col. Guy M. Townsend Air Force test pilot; Captain John Elrod, the Bombing-Navigational System project officer, and the amazing Wellwood E. Beall, Boeing's senior vice president at the time. Beall was a brilliant engineer and a brilliant salesman. (Boeing)
In the second picture, with no General around, Art Curran, Townsend and Tex Johnson grin it up in front of XB-52. (Boeing)

TESTING AND INITIAL PRODUCTION

Roll out and First Flight

The experience Boeing had gained in its years of bomber production stood it in good stead, and the XB-52 and YB-52 aircraft rapidly took shape in a highly classified area of the Seattle plant.

Contrary to previous accounts, there was basically no difference between the two aircraft. The designation was changed to YB on the second aircraft solely to permit the Air Force to expend an additional $10,000,000 in production funds on the experimental aircraft.

During the period of actual construction of the two aircraft, Boeing was experiencing some real problems with its suppliers. The electrical, pneumatic and hydraulic systems all called for innovation on the part of subcontractors. In many cases the Boeing Company itself provided engineers to work directly with suppliers to see that requirements were met. The multiple sleeve type fuel cells were a typical example. Unlike the bag types of the past, these had to be sealed on each end of the sleeve, with the wing inspar ribs serving in effect as dividers. Conventional thinking at the rubber companies which had previously manufactured the bag type had to be overcome by some intensive Boeing persuasions before the new tanks were ready.

One factor which helped the success of the B-52 programme was that (as previously noted) Art Carlsen, the project engineer, had had a large part in the original weight estimates. Carlsen, who is perhaps the unsung hero of the B-52 programme, was miserly with every pound, and he insisted that both Boeing and its suppliers somehow keep within the original weight allocations. When the aircraft was rolled out, it was slightly under the original predicted empty weight, a most unusual situation.

Carlsen was later succeeded as project engineer by a young man named T. A. Wilson. He did a fine job, and was subsequently selected by Ed Wells to head the Minuteman programme. Eventually "T" as he was called became first president and then chairman of the board of the Boeing Company.

The first two B-52s looked, on paper, to be very similar to the familiar B-47. As they took shape in the factory, however, it was clear to everyone that they were very different, much more sophisticated aircraft. The long wings stretched out for 185 feet, and when not loaded with fuel were high above the ground so that not even the tip protection gear touched. The prototype Pratt & Whitney YJ57-P-3 engines were housed in

Below: *The YB-52 on takeoff roll. Contrary to published reports, the XB and YB aircraft were identical except that the YB was instrumented for flutter tests and the XB was not. YB flew first because XB had encountered damage during full pressure test of pneumatic system.* (Boeing)

pairs in nacelles, and the slender canopy looked almost tiny on the fuselage nose.

The XB-52 was rolled out of the factory in great secrecy on the night of November 29, 1951. Shrouded with tarpaulins it was moved to the flight test hangar for system and taxi testing. During preparation for flight, the XB-52 underwent one final test of its pneumatic system. At full pressure, the system suffered a massive failure, ripping out the whole trailing edge of the wing. It was secretly returned to the construction area for what was officially termed "equipment installation", and as a result the YB-52, which was rolled out on March 15, 1952, made the first flight for the type on April 15, 1952. The XB-52 did not make its first flight until October 2, 1952.

The gigantic aircraft was exploring new ground, and Paul Higgins and Vaughn Blumenthal, the aerodynamicists, were worried about the possibility that aileron overbalance might be encountered. As a result, they made the control forces very high for the first flight of the YB-52, and set the pick-up point of the spoilers, which augmented aileron control, at about 45 degrees of control wheel movement.

A. M. "Tex" Johnson—one feels almost obligated to say "the legendary Tex Johnson"—and Lt Col Guy M. Townsend were the first flight pilots. Boeing and the Air Force had agreed that the usual practice of having company flight tests followed by Air Force flight tests was too expensive, and Townsend, who had done so much to make the B-47 programme a success, had the complete confidence of both organizations.

The first flight was from Boeing Field, in what was then South Seattle to Moses Lake, Washington. The take-off and flight over were without incident except that the lateral control forces, as Blumenthal and Higgins had arranged, were very high. When the pilots landed their enthusiasm was guarded. One of the engineers asked what Johnson thought was needed, and Johnson replied, dead pan, "New flight suits". Somewhat nonplussed, the engineer said "New flight suits? I thought you had new flight suits?" Johnson snapped

Above: *The XB-52, showing the thickened wing root section that made such a difference to its weight and fuel capacity. The decision was made long before production of the B-52A to go to side by side seating, for purposes of crew coordination.* (Boeing)

Left: *Famed Boeing test pilot "Tex" Johnson and Lt. General Donald Putt next to the YB-52. Johnson shepherded many Boeing aircraft through initial flight test programmes, and made a name for himself when he rolled the prototype 707 (the Dash-80) not once but twice on its initial public display in Seattle. Putt was a tremendously important figure in Air Force development activities.* (Boeing)

Top right: *Boeing was determined to give the B-52 enough wing area; all 4,000 square feet shows up well here. Shorter nose of X and Y aircraft is apparent in this angle.* (Boeing)

Centre right: *Another great improvement over the B-47 was use of "all flying stabilizer". Note relative degree of available movement shown by darker area at leading edge of horizontal stabilizer. Trim was very important in operating the B-52, because elevators were so small.* (Boeing)

Right: *The first photograph released showing the B-52's complex gear, which had been kept highly classified. Gear is coming up; odd angles result from the natural retraction geometry.* (Boeing)

61

back "If we are going to have to man-handle this son-of-a-bitch around, we're going to have arms bigger than our legs and we'll need new flight suits."

The re-rigging of the aircraft was easy, however, and subsequent flights were a pleasure. The Boeing team had come to Moses Lake anticipating a period of several weeks to iron out problems so that the aircraft would be safe to fly over the populated Seattle area. So right was the aircraft, however, that after one week it was flown back to the plant to continue tests.

One happy, unexpected test result was that the drag of the XB and YB-52 was 11% lower than predicted. The traditional Boeing conservatism had resulted in better than planned performance just as it had in the B-47.

The two prototypes were subjected to a long series of development programmes, and were ultimately intended to go to the Air Force Museum in Dayton, Ohio. Unfortunately they were caught in a base beautification programme, a spin-off of the national drive led by President Johnson's wife, Lady Bird, and were cut up for scrap.

Top left: *Posed shots show crew in pressure suit regalia. Flight tests showed that the B-52's normal performance envelope would not call for much flying at altitudes in excess of 50,000; as pressurized suits were only required for above 50,000, they were soon abandoned for normal operations.* (Boeing)

Above left: *Radar Navigator and Navigator positions. Crews gladly gave up suits which were cumbersome, uncomfortable and difficult to get into.* (Boeing)

Above: *The Electronic Warfare Officer's position has perhaps changed the most radically in equipment if not in appearance, as the requirement for defensive electronic countermeasures grew.* (Boeing)

Right: *Only three B-52As were built, the first flying on August 5, 1954. Most noticeable change from the YB-52 illustrated is the lengthened nose section with its tandem seating.*

The Production Aircraft

Boeing received a letter contract for 13 B-52As in February, 1951. These production aircraft were subject to the greatest configuration change until the advent of the B-52Gs. As mentioned previously, the tandem cockpit was changed at LeMay's request to the now familiar side by side type, and the fuselage was lengthened by four feet. Improved J57-P-9W engines were installed and 1,000 gallon external jettisonable tanks were fitted outboard on the wings. These provided additional fuel and also gave load relief to the structure. Test flights revealed that the A series drag was again less than predicted, this time by 2.5%.

What corresponded to the old service test order for 13 aircraft had been placed for the B-52A, and even though the somewhat doubtful competition of the Convair YB-60 (essentially a swept wing J57 powered version of the B-36, without any significant improvement of aerofoil or structure) still hung over their heads, the Boeing engineers were already at work on more advanced models of the aircraft.

The first production B-52A flew on August 5, 1954; only a few days later a contract was let for 50 B and RB-52Bs, and by December 1954, the plant at Wichita, Kansas was confirmed as a second production source.

The order for 13 B-52As was subsequently cut to three, with the remaining ten becoming B-52Bs. The three A models entered development and after serving as test beds for B-52 systems did a variety of work in other capacities. The first in the series, appropriately serialled 52-001, was redesignated NB-52A and served as a launch vehicle for the North American X-15 rocket powered research workhorse. (The "N" in NB indicated that the aircraft was too highly modified to be returned to operational configuration for squadron use.)

Enthusiasm for the B-52 grew, and despite the high unit costs attributed to the first three aircraft (some $29,000,000 each, which reflected amortized preproduction costs) orders were placed in rapid fashion, and unit costs fell. The accompanying table, 4-1, depicts the costs for the aircraft, spares and other equipment, and shows that unit costs actually dipped to $3.7 million in 1959.

Average Unit Costs of B-52 Aircraft

TMS	Airframe	Installed Engines	Electronics	Ordnance	Other Including Armament	Total
B-52A	26,433,518	2,842,120	50,761	9,193	47,874	29,383,466
B-52B	11,328,398	2,547,472	61,198	11,520	482,284	14,430,872
B-52C	5,359,017	1,513,220	71,397	10,983	293,346	7,247,963
B-52D	4,654,494	1,291,415	68,613	17,928	548,353	6,580,803
B-52E	3,700,750	1,256,516	54,933	4,626	931,665	5,948,490
B-52F	3,772,247	1,787,191	60,111	3,016	862,839	6,485,404
B-52G	5,351,819	1,427,511	66,374	6,809	840,000	7,692,913
B-52H	6,076,157	1,640,373	61,020	6,804	1,501,422	9,285,776

Source: T.O. 00-25-30

Above: *An RB-52B, one of twenty seven built with reconnaissance capability.* (U.S. Air Force)

The B-52B was outwardly identical to the A model, but had an increased reconnaissance capability and the MA-6A bombing navigation system, a vast improvement over what had been previously considered revolutionary in the B-47, the K system.

Gross weight for both the A and B models was increased to 420,000 pounds, up from the "unofficial" 405,000 pounds of the XB and YB-52.

Twenty-three B-52Bs used later models of the J57 engine, including the J57-P-19W, J57-P-29W and J57-P-29WA types. The -19W had an increased thrust rating of 12,100 pounds with water injection, and 10,500 pounds without, and used a titanium compressor. Pratt & Whitney built 547 and Ford built 1,416 of the -19Ws. The -29W had a steel low compressor and titanium high compressor, and delivered 11,000 pounds of thrust with water injection; the -29WA had twice the water flow of the -29W and a 12,100 pound static thrust rating.

Twenty-seven of the series became RB-52Bs, using a two man pressurized capsule in the bomb bay which could perform electronic countermeasure ferret work, or photographic reconnaissance. The two men who climbed into the capsule which was tucked into the cavernous bomb bay were of a special breed, for theirs was a dark and remote world, cut off from all but radio communication, and extremely vulnerable to any sort of low altitude emergency where their downward ejection seats would be at a disadvantage.

The first B-52B, 52-8711, was delivered with a flourish to Castle Air Force Base, Merced, California, on June 29, 1955. Within a few months the 93rd Bomb Wing converted from B-47s to B-52s without losing its nuclear capability, and set up the 4017th Combat Crew Training Squadron which became the focal point for all B-52 training for the next several years.

Above: Castle Air Force base served as the base not only for crew training, but for the initial operational introduction of the B-52B regarding maintenance, armament, and so on. Much special equipment was needed for the huge aircraft. (U.S. Air Force)

Left: The pod being installed in an RB-52B. Thus equipped the aircraft could be used either for photographic or electronic reconnaissance. (Boeing)

Below: The 24,325 statute mile flight round the world was made during January 16 to 18, 1957. Flight took 45 hours and 19 minutes, and was within two minutes of the expected time of arrival. Fuel consumption had been computed within ½ of 1% of actual. Two aircraft had no maintenance write-ups, the third had to have an alternator replaced. (Air Force)

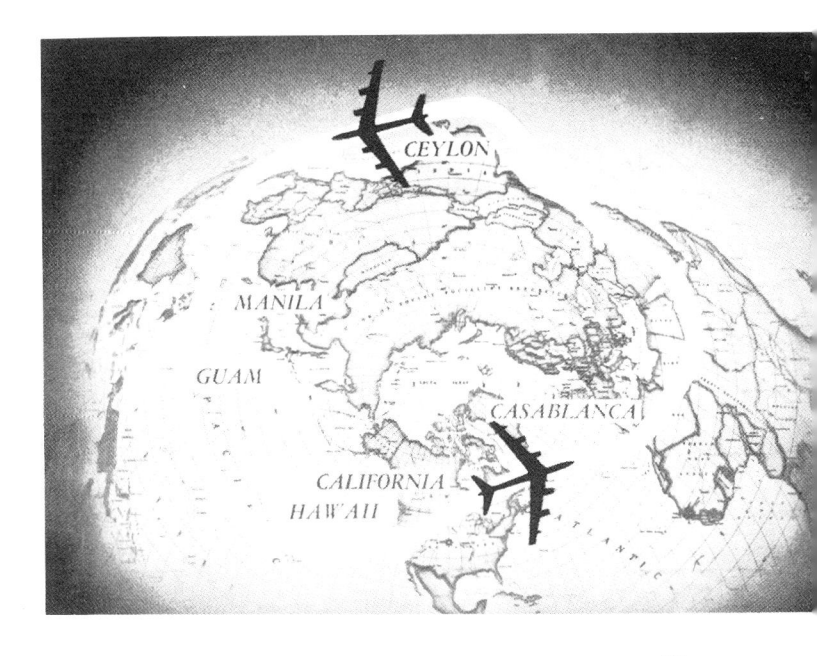

Fuel leaks caused some serious problems during the first year, but in general all of the training requirements were met. Transition had proceeded so well that the 93rd won the 1956 Strategic Air Command Annual Bombing and Navigation Competition, a prestigious "world series" of bomber aircraft trials. It was the first time ever that a new aircraft had won in its initial appearance, and was a tribute to the superb professionalism of Castle crews.

The striking power of the new unit was demonstrated in January 1957, when three aircraft, commanded by Major General Archie Olds, flew non-stop around the world.

Five aircraft, including two spares, took off at 1300 PST on January 16; from Castle they flew via Newfoundland, Casablanca, Dhahran, Ceylon, the Malay Peninsula, Manila and Guam, returning to Castle only to find that the typical San Joaquin valley winter weather had forced a diversion to March Air Force Base in San Bernadino, California. One spare, piloted by Guy Townsend, had to divert to Goose Bay when the inflight refuelling receptacle iced over; the second made a planned landing in England.

The 24,235 mile flight was completed in 45 hours and 19 minutes, which was within two minutes of the predicted time en route; the fuel consumption was within ·5% of predicted. All 24 engines were still running, the bomb/nav systems, which had operated continuously, were all in excellent condition. After the flight two of the aircraft required no maintenance prior to the flight back to Castle, while one had to have an alternator replaced.

The crews had been carefully selected for the flight, and

had been augmented by an additional pilot and navigator so that an eight hour on, four hour off shift series could be maintained. The pilots were specially selected for their skill in inflight refuelling, while the navigators were all qualified as instructors.

The idea for the flight had originated with Colonel Don Hillman, and Castle was selected as the unit because its alternators were of the liquid lubrication type and could be modified for the extremely long duration of the flight. Other than this the aircraft were stock. A hammock was installed in the lower compartment and a collapsible plywood bunk installed behind the pilot's seat. A new delicacy appeared on the flight, one which would find its way into Air Force folklore, the bite-size steak, wrapped in tin foil and frozen, and which could be heated in the B-3 oven.

Left: *General LeMay shown pinning the Distinguished Flying Cross on Major General Archie Olds; each man on crew was decorated with DFC. Captain Dingwell, fourth from right in second rank, is undoubtedly wishing that he had a cap on, for LeMay would undoubtedly take notice.* (U.S. Air Force)

Below: *The first B-52C flew on March 9th, 1956; it was the first to have the anti-nuclear blast white paint and 3,000 gallon underwing tanks. Gross weight had climbed to 450,000 pounds. The B-52C retained the B model's multi-mission capability.* (National Air and Space Museum)

One nostalgic side note was that the aircraft commander of the lead aircraft, Lieutenant Colonel James H. Morris, had been co-pilot on the "Lucky Lady II", the B-50D which had made the first non-stop trip around the world in 1949 in 94 hours and 10 minutes.

When the 93rd Bomb Wing had completed conversion to the B-52, other SAC wings began feeding in to Castle, starting the conversion process which quickly changed SAC from a B-47 to a B-52 orientation.

Aircraft updating continued. The first of 35 B-52C aircraft made its first flight on March 9, 1956. These aircraft typified the concurrent process of development, contracting, production and delivery. The design had been initiated in December, 1953, the first aircraft entered final assembly in October, 1955, and the first acceptance flight took place in June, 1956.

Essentially similar to the B-52B, some of these carried an internal Boeing designation of RB-52C, but were referred to only as B-52Cs in the field, even though they retained the reconnaissance capability.

The latest model was distinguishable on sight for two reasons. It was the first to be painted white on the undersurfaces to reflect thermal heat from an atomic detonation, and it also incorporated huge 3,000 gallon drop tanks. The gross weight was increased to 450,000 pounds, with a total fuel capacity of 41,550 gallons and a military load capacity of 64,000 pounds. Some Cs also had the MD-9 fire control system, an improvement over the earlier A-3A system.

Even as production programmes overlapped, so did the modification programmes which will be covered in Chapter Seven.

The decision to shift B-52 production from Seattle to Wichita was conditioned upon a number of factors. The Wichita plant's production of B-47s would be coming to an end, and there was a large, well qualified work force on hand. (Boeing has always drawn the cream of the large crop of qualified aircraft workers in the Beech-Cessna-Boeing town, by the simple device of paying slightly higher wages.) The commercial business at the Seattle plant was picking up, and the pearl of the Northwest would not be able to accommodate the expanding civil and military business without forcing a boom and bust sort of expansion on the local economy. (This in fact happened anyway as a result of the unprecedentedly good commercial sales in the years 1966-68. Boeing, despite its best efforts, found itself in a recession in the early 1970s, and had to cut its work force back drastically.)

Seattle began to taper off its B-52 production with the B-52D, of which 101 were built there, while 69 others were built at Wichita. The trend continued, with Wichita building 58 of 100 B-52Es, 45 of 89 B-52Fs, and all of the 193 B-52Gs and 102 B-52Hs. When the total of 744 B-52s had been completed, Wichita had built 467, and Seattle 277.

Wichita then embarked upon a continuous maintenance and modification programme to keep the fleet effective. Seattle, meantime, had focused more exclusively on commercial aircraft production, while keeping its hand in missile work, and maintaining C/KC-135 tanker/transport production.

The Size of the Matrix

The smooth transition to the Wichita facility characterized the almost incredible harmony within the company and within the Air Force, and between the company and the Air Force. During this period, when only about 35% of the aircraft had been completed, two enormous manufacturing centres had been set up as well as two huge maintenance centres, the Oklahoma City Air Material Area (OCAMA) at Tinker Air Force Base, Oklahoma, and the San Antonio Air Material Area (SAAMA) at Kelly Air Force Base, Texas. (In AFLC parlance, these are called "Brown eyed" centres because of their loyal capable Hispanic-American work force.)

An important adjunct to the B-52 programme came into being when the versatile, long-lived KC-135A tanker made its first flight on August 31, 1956. Definitely *not,* as is so often assumed, a military version of the commercial transport, the KC-135A was developed in parallel with the 707, and both were derived from the original 367-80 prototype. Not only do the two types share only a common wing box and the overall configuration, their designs were predicated on two different philosophies—the KC-135 being designed to a "safe-life" to obtain a better strength-to-weight ratio, while the 707 was designed to meet FAA "fail-safe" standards.

The KC-135A has had the longest production run in history as a single model, while its modifications have reached further than any other model, attaining the KC-135W designation already, and very possibly in the future reaching a double alphabet suffix.

This is not the place for a history of the KC-135A programme, but it is important to mention that it was developed concurrently with both the B-52 and commercial transports, and illustrates the Boeing Company's ability to expand without diluting its management or compromising its product. In its sphere, the KC-135A has an equally important weapon system as the B-52, supporting as it does the refuelling requirements of SAC, the Tactical Air Command, the Military Airlift Command, Air National Guard units, plus NATO, Allied and some Navy commitments. SAC was designated the single manager for KC-135 tankers in 1961.

The expansive Wichita facility actually built the first B-52D, rolling it out on December 7, 1955 and achieving a first flight on May 14, 1956. Seattle's first D model flew on September 28 of the same year. The 170 D models became the backbone of the fleet for several years, and then had a new lease of life as "iron bomb" carriers during the Viet Nam conflict. Originally not much different than the B-52Cs, the Ds were the recipients of many changes to enable them to alter their roles from high altitude nuclear bomber to a high level dropper of conventional weapons.

Below and right: The first production B-52D at Wichita, "clean" and "dirty". The drag from all of the flap and gear and spoilers hanging out slows the B-52 down rapidly, and makes control of approach speeds much easier. (Boeing)

The B-52Es were the first to use the new low altitude equipment which was then deemed to be necessary to elude the ever expanding Soviet radar and missile network. The deployment of the Russian air defence system, when depicted upon a map of the USSR, almost defies belief. The entire border of the country is encrusted with overlapping radar defences, and dotted with anti-aircraft, ground-to-air missiles and interceptor units. The defence is in depth, stretching inland and blossoming forth again around the approaches to major cities. The cost of the system is incalculable, but it reduced Soviet vulnerability to high altitude bomber penetration to vanishing point, and leaves open still only the narrowest corridor above the earth for terrain following aircraft to creep under. Even this gap is being narrowed day by day with improved radar systems, "look-down" airborne radars, and ever "smarter" missiles.

Above: The operations in Viet Nam required that the B-52D acquire a much more sinister paint job. Now outfitted for conventional weapons delivery, the reflective white paint was no longer needed. (Courtesy Jay Miller)

Right: Close up of rear gunner's position and four machine gun turret. Long tubular "horns" on either side of gunner's radar dome are ECM antennae. Turret jettisons just along line shown at edge of elevator. (Courtesy Jay Miller)

Top: *The B-52D was modified to carry one hundred and eight 500 pound bombs for operations in Viet Nam. Although less modern than the B-52Gs, it had a better ECM capability, and was preferred because the tail gunner could monitor SAM launches from the rear. (Boeing)*

Above: *One hundred B-52Es were purchased; they had uprated bombing and navigation equipment. (Boeing)*

Left: *The B-52F was the last to be built in Seattle; it had the uprated J57-P-43W engine of 13,750 pounds of thrust. Wichita built 45 of the Fs to 44 for Seattle. This aircraft is operating from Minot Air Force Base, Wyoming; crews used to despair when they learned they were being assigned to the cold Northwestern base, but most soon grew to love the area. (U.S. Air Force)*

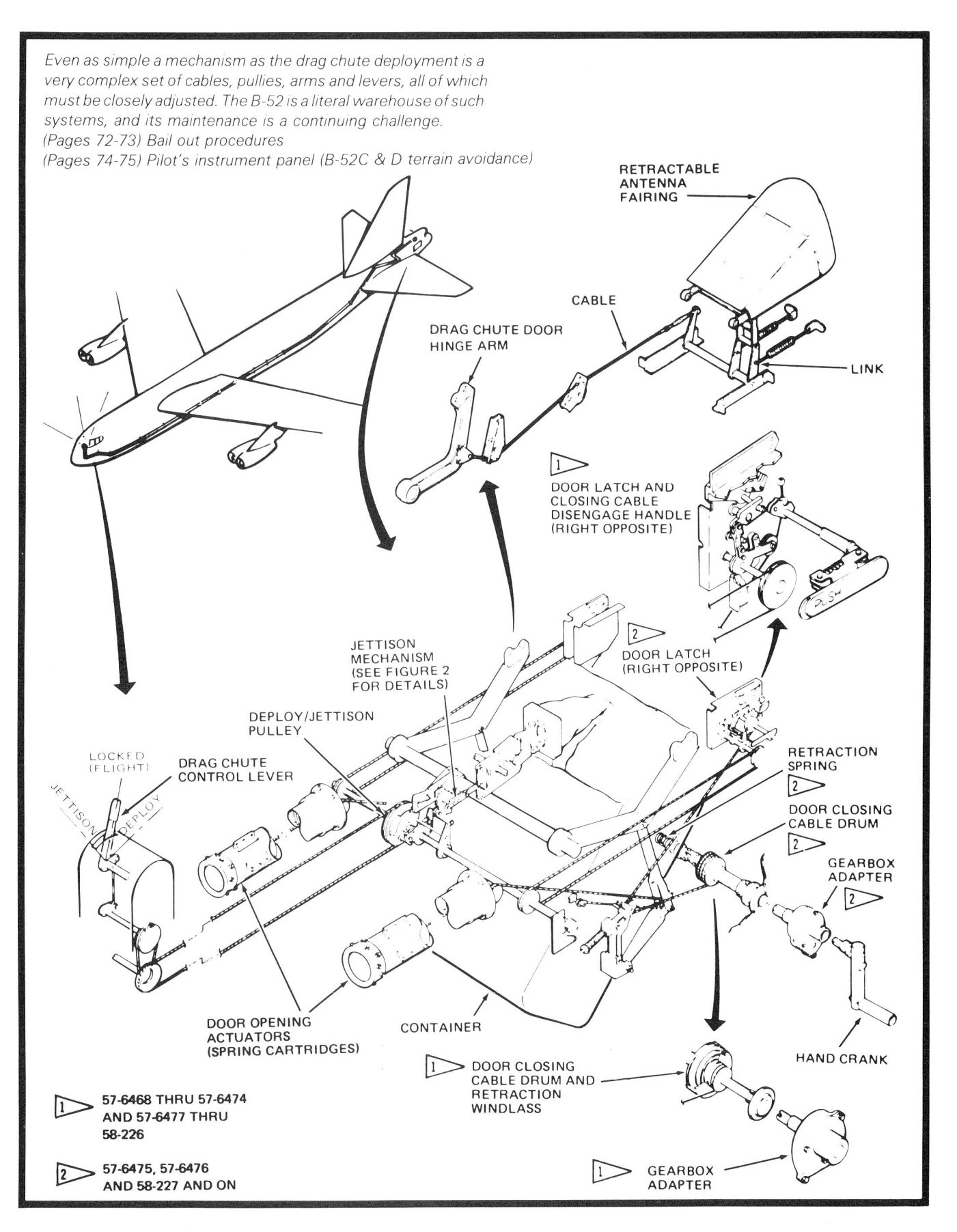

Even as simple a mechanism as the drag chute deployment is a very complex set of cables, pullies, arms and levers, all of which must be closely adjusted. The B-52 is a literal warehouse of such systems, and its maintenance is a continuing challenge.
(Pages 72-73) Bail out procedures
(Pages 74-75) Pilot's instrument panel (B-52C & D terrain avoidance)

UPWARD EJECTION

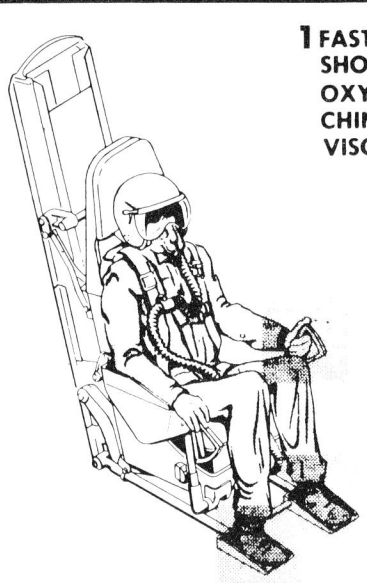

1. FASTEN SAFETY BELT, SHOULDER HARNESS, OXYGEN MASK AND CHIN STRAP. HELMET VISOR DOWN
2. PLACE FEET IN FOOT RESTS
3. RAISE ARMRESTS
4. ROTATE LEFT ARMING LEVER UPWARD
5. HEAD AGAINST HEAD REST

NOTE
IF HATCH DOES NOT JETTISON, PROCEED TO ALTERNATE EXIT.

6. ROTATE RIGHT ARMING LEVER UPWARD
7. SQUEEZE TRIGGER TO FIRE SEAT 

DOWNWARD EJECTION

1. FASTEN SAFETY BELT, SHOULDER HARNESS, OXYGEN MASK AND CHIN STRAP. HELMET VISOR DOWN
2. PLACE LEGS IN POSITION AGAINST ANKLE RESTRAINT TRIGGERS
3. HEAD TIGHT AGAINST HEAD REST 
4. ROTATE ARMING LEVER UPWARD AND FORWARD Less KQ

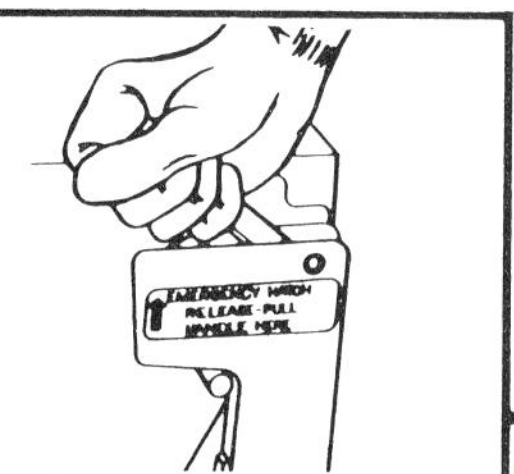

IF HATCH DOES NOT JETTISON - PULL HATCH RELEASE

5. GRASP TRIGGER RING WITH BOTH HANDS, HOLD ELBOWS TIGHTLY AGAINST BODY AND PULL RING TO FIRE SEAT

BAILOUT FROM NAVIGATOR'S ESCAPE HATCH

GEAR UP-275 KNOTS (IAS) OR LESS
GEAR DOWN-250 KNOTS (IAS) OR LESS

BAILOUT FROM RIGHT REAR WHEEL WELL

AIRSPEED-275 KNOTS (IAS) OR LESS

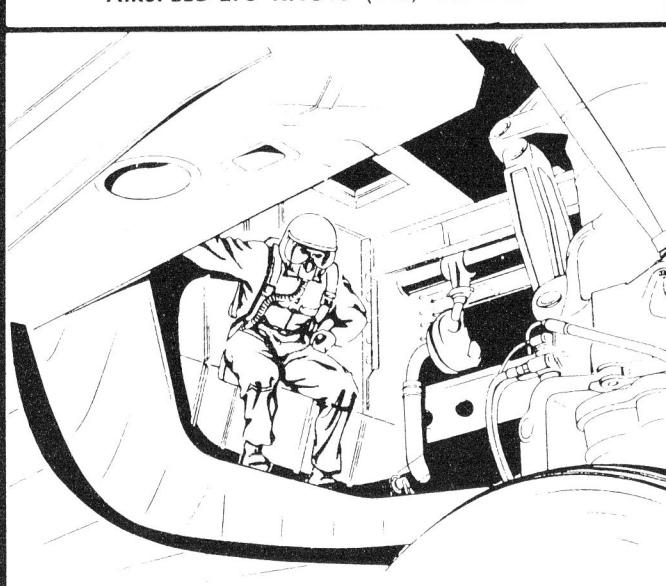

IF TURRET DOES NOT JETTISON REQUEST PILOT TO LOWER RIGHT AFT LANDING GEAR — FOLLOW ALTERNATE BAILOUT PROCEDURE

GUNNER'S BAILOUT PROCEDURE

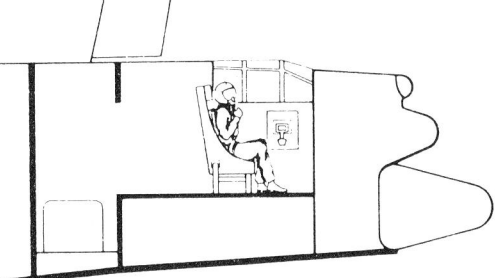

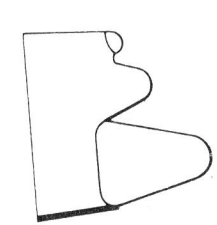

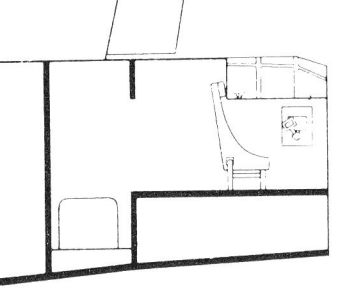

1. FASTEN SAFETY BELT, SHOULDER HARNESS, OXYGEN MASK AND CHIN STRAP. HELMET VISOR DOWN
2. PULL UP TURRET-DRAG CHUTE INTERCONNECT KNOB UPON PILOT COMMAND ONLY
3. PULL JETTISON HANDLE ON PILOT'S COMMAND • NOTIFY PILOT READY TO BAILOUT
4. PULL BAILOUT BOTTLE RELEASE CORD
5. UNFASTEN SAFETY BELT

WARNING

Pull integrated harness release handle only when using modified B-5 parachute.

6. LEAVE THE AIRPLANE WITH ARMS AND LEGS HELD CLOSE TO BODY
7. PULL PARACHUTE ARMING LANYARD KNOB Less [LJ]

BAILOUT PROCEDURES

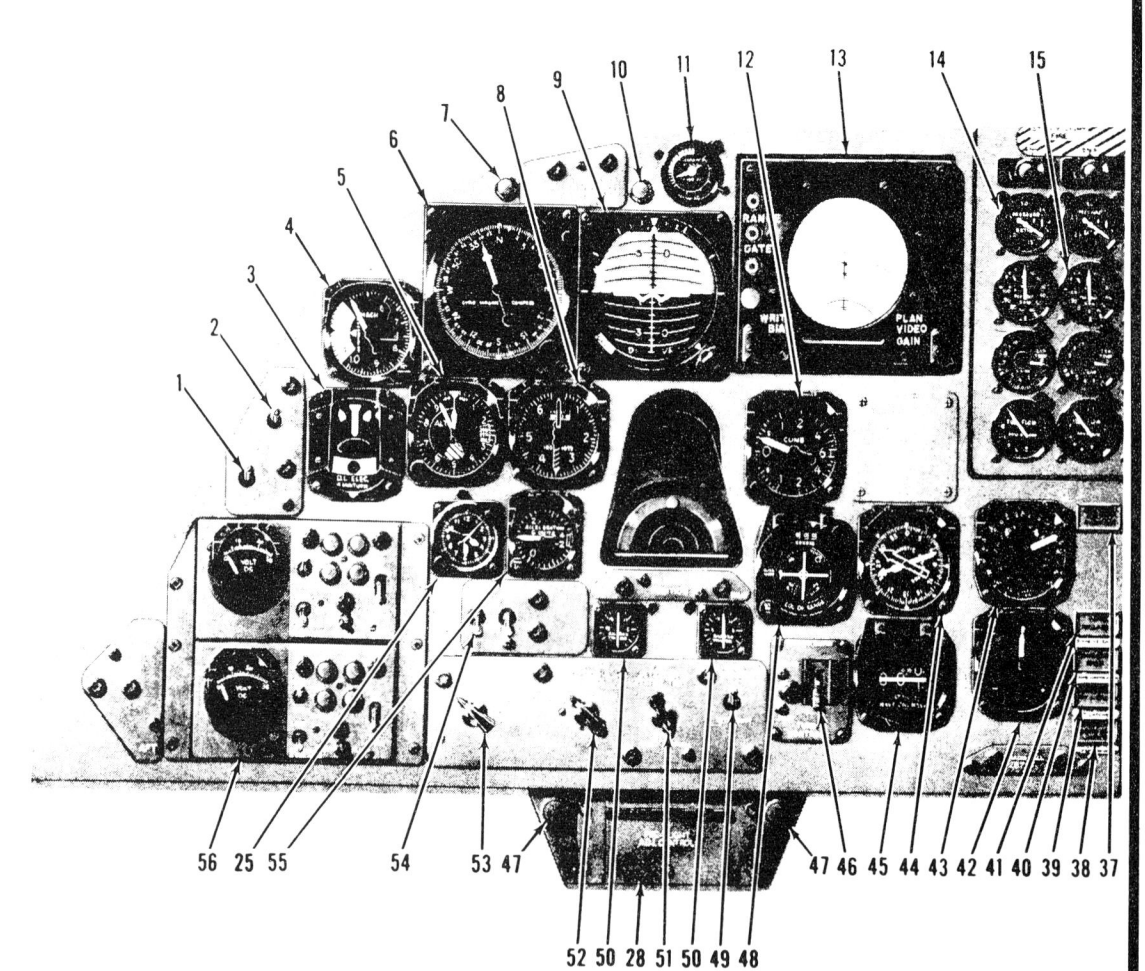

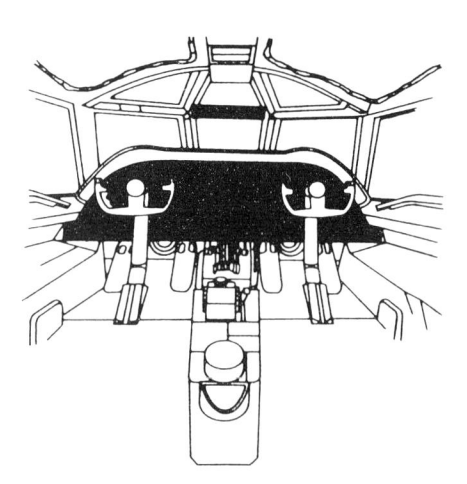

1. BOMB DOORS SWITCH
2. MACH INDICATOR SWITCH
3. TURN-AND-SLIP INDICATOR
4. MACHMETER
5. ALTIMETER
6. DIRECTIONAL INDICATOR (N-1 REPEATER)
7. AUTOMATIC PILOT DISENGAGED LIGHT
8. AIRSPEED INDICATOR
9. PILOTS' ATTITUDE INDICATOR
10. HYDRAULIC PACK PRESSURE LOW MASTER LIGHT
11. CLEARANCE PLANE INDICATOR
12. VERTICAL VELOCITY INDICATOR
13. TERRAIN DISPLAY INDICATOR
14. ENGINE PRESSURE RATIO GAGES
15. TACHOMETERS
16. EXHAUST GAS TEMPERATURE GAGES
17. MAGNETIC STANDBY COMPASS
18. OIL PRESSURE GAGES

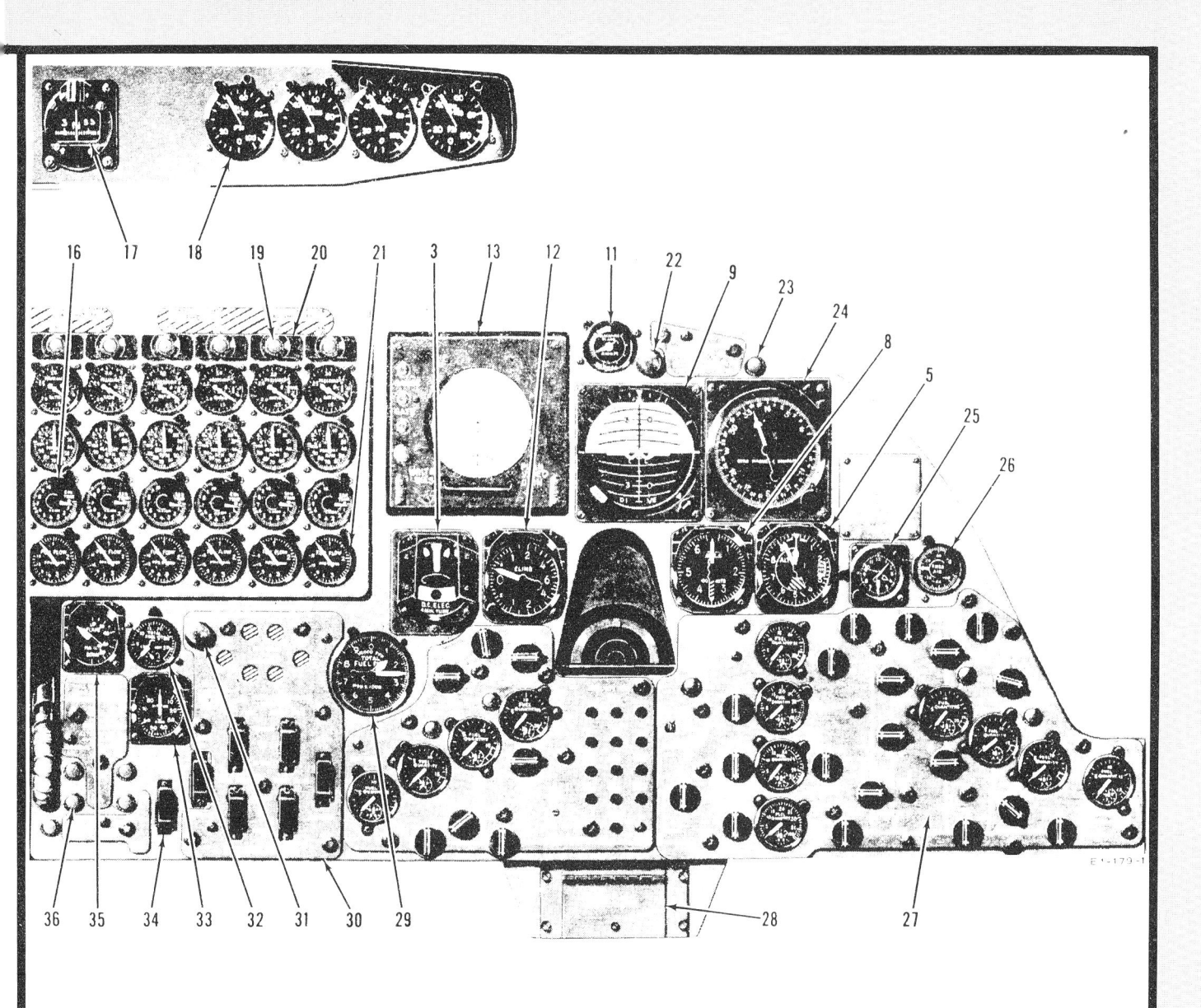

19.	ENGINE FIRE WARNING LIGHTS	38.	HATCHES NOT CLOSED AND LOCKED LIGHT
20.	FIREWALL FUEL SHUTOFF SWITCHES	39.	BOMB DOORS NOT LATCHED LIGHT
21.	FUEL FLOWMETERS	40.	BOMB DOORS OPEN LIGHT
22.	MASTER FUSELAGE OVERHEAT (FIRE) WARNING LIGHT	41.	BOMB RELEASED LIGHT
23.	GUNNER'S CABIN PRESSURE WARNING LIGHT	42.	LATERAL ERROR METER NH
24.	DIRECTIONAL INDICATOR (GYRO)	43.	PILOT'S DATA INDICATOR (PDI)
25.	CLOCK	44.	RADIO MAGNETIC INDICATOR
26.	OUTSIDE AIR TEMPERATURE GAGE	45.	DISTANCE INDICATOR
27.	FUEL SYSTEM CONTROLS (FIGURE 1-16)	46.	AUTOPILOT TURN CONTROL SELECTOR SWITCH
28.	TERRAIN PREFLIGHT ADJUST CONTROL	47.	AIR OUTLET KNOB
29.	TOTAL FUEL FLOW INDICATOR	48.	OMNI-RANGE RADIO COURSE INDICATOR
30.	LANDING GEAR CONTROLS (FIGURE 1-41)	49.	TONE SCORING INTERRUPT SWITCH
31.	STORE JETTISONED LIGHT NH	50.	AILERON TRIM INDICATOR
32.	TOTAL FUEL QUANTITY GAGE	51.	ENGINE FIRE DETECTOR SYSTEM TEST SWITCH
33.	TAIL COMPARTMENT ALTIMETER	52.	WINDSHIELD ANTI-ICE AND DEFOGGING SWITCH
34.	ANTISKID SWITCH	53.	WINDSHIELD WIPER SWITCH
35.	WING FLAP POSITION INDICATOR	54.	PITOT HEAT SWITCHES
36.	ALTERNATOR OVERLOAD LIGHTS	55.	ACCELEROMETER
37.	ANTI-ICING SURFACE OVERHEAT LIGHT	56.	T-18 CONTROL PANELS

PILOTS' INSTRUMENT PANEL NQ

The B-52Es had an improved bombing navigation capability, using the AN/ASQ-38 integrated system which took data from many auxiliary sources, including the astrocompass and the automatic navigation radar system. The E model was also the first to test the Hound Dog missile, the first thermonuclear air-to-ground missile adopted by SAC, and which was subsequently fitted to the G and H models.

One of the hard early Boeing decisions, the use of air drive turbines for electrical generating systems to save weight, had caused some tragic problems. The turbine wheels, driven at very high RPM, had on one occasion disintegrated, sending red hot fragments into the fuselage fuel cells, creating a catastrophic fire. The short term solution had been a change to fail-safe turbine wheels, but the long term solution was the adoption of "hard drive" alternators, mechanically driven by the left engine in each of the four engine pods. The new method caused some extra weight and drag, but was more powerful and reliable, and eliminated the turbine wheel disintegration hazard. Boeing and Sundstrand worked closely to achieve the new constant speed drive for use on the E and subsequent models.

The next in the series, the B-52F, made its first flight on May 6, 1958, and was the beneficiary of numerous improvements, the most notable being the incorporation of J57-P-43W engines. These offered additional performance, with 13,750 pounds of thrust, which in turn introduced stress in some of the secondary structures such as the ribs of the wing trailing edges. P-43W engines were installed on a B-52C airplane, and ground runs at full power with water injection were conducted on a continuous basis to determine and cure sonic fatigue problems. More than 1,000 secondary structural changes of this type were incorporated in the B-52F.

By the time the last B-52F had made its delivery flight from Seattle on February 25, 1959, the SAC fleet had reached an almost intimidating strength. As early as 1956, however, the need for greater range and flexibility had been seen, and production of the B-52G had been authorized in August of that year. Outwardly the G resembled its predecessors except for a shorter vertical surface and a lack of ailerons. Internally, however, it was an almost entirely new aircraft, one in which Boeing would pull off the famous engineering hat trick of lowering basic empty operating weight while increasing gross operating weight. There were, however, some problems still waiting to be discovered and solved.

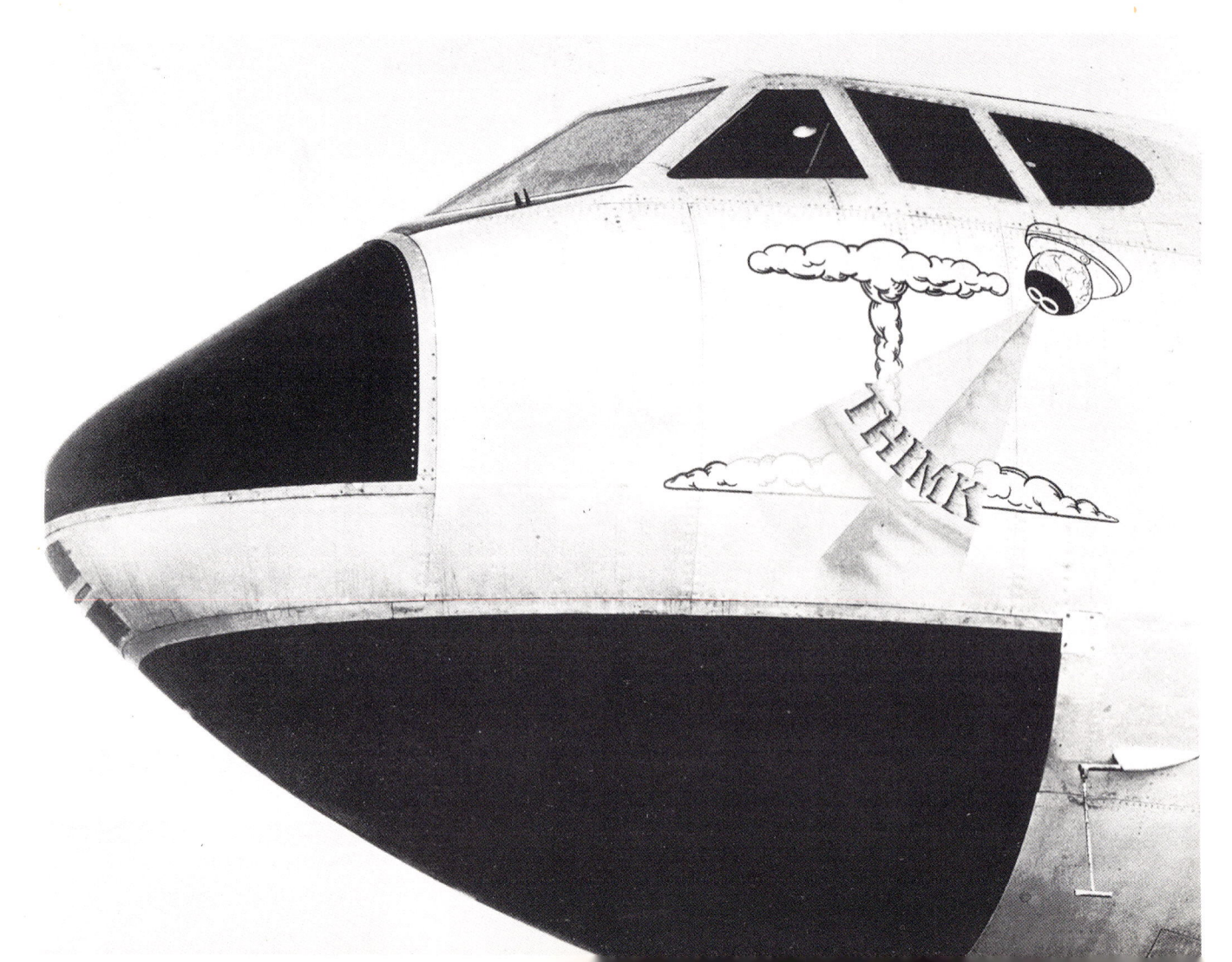

Below: The International Business Machine designed computer was the heart of the MA-2 bombing system; considerable liberty was taken with the spelling of the traditional IBM motto "Think" as well as with the use of an 8-Ball eye. Some sardonic maintenance type was probably taking a little revenge on the BRANE system.

GILDING THE LILY – THE Gs AND Hs

In March 1956, Boeing made a presentation to the Air Force for a major improvement in the B-52. The briefing offered some tempting possibilities: an increase in range of 30%, a decrease in maintenance man-hours of the order of 25%, a decrease in operating weight empty of 15,000 pounds, and an increase in electronic warfare capability of 70%.

The actual design improvements included a conversion from bladder cells to an integral fuel tank, a "wet wing" which called for a complete structural redesign and a weight saving of 5,847 pounds. The new wing required the machining of long alloy wing skins so that stiffeners were an integral part of the structure, and resulting in a surface with a minimum of chordwise joints. This reduced the possibility of fuel leaks and fatigue, and incidentally tied up a great proportion of US machining capability at the time. The vertical fin height was reduced from 48 feet to 40 feet, and this, with the elimination of the aileron system, saved another 12,000 pounds.

A new fire control system allowed the gunner to be moved from his lonely—and heavy—rear turret to the front compartment, where he could operate the armament by a closed circuit television or radar. Visibility for inflight refuelling was enhanced by lowering the crew deck by two inches, a small change with really dramatic effect.

Perhaps even more significant, in the light of later strategic developments, the B-52G was designed as a missile platform as well as a gravity bomber. The supersonic GAM-77 Hound Dog was produced for use with this aircraft, which could also have accommodated the Skybolt air launched ballistic missile if it had been procured.

The external clues identifying the G model included a one foot longer fuselage, a shortened one piece nose radome, use of spoilers only for lateral control, remote gunner, smaller external fuel tanks and revised nacelles. Internally, many new systems provided both the problems and the potential of a brand new aircraft. The major development responsibility was given to Boeing, Wichita, which became the sole source of the aircraft.

The wing structure was not only lighter, but had to carry the higher gross weight of 488,000 pounds specified for the aircraft. In addition to being strong and extraordinarily flexible, it had to be leak proof under conditions of repeated flexing, hard landings, gusty air, and so on. A new trailing edge structure was designed, once again, to offset sonic fatigue.

The wing skin and stiffeners, masterpieces of numerically controlled milling, lacked durability and had to be reinforced on two separate occasions. As we shall see, there would be other problems.

The new fuel tanks required an entirely new fuel management system, including employing fuel usage sequencing for centre of gravity control. Smaller 700 gallon external tanks were fitted, not so much for additional fuel capacity as to act as fuel filled bob weights to help prevent flutter. The inboard nacelle struts were modified for increased dynamic stability in flight.

Total internal tankage was 46,575 gallons; with the two external tanks, fuel capacity was 47,975 gallons.

It was found that with the shorter fin and the lack of ailerons, the aircraft had an increased tendency to Dutch Roll. This motion had been the bane of all large swept wing aircraft and the existing yaw damper was not adequate.

Besides the structural changes, there was a host of equipment changes, most of which delighted the SAC crews once they had adapted to the change in flight characteristics. The elimination of ailerons changed the lateral response; when a turn was initiated, the spoilers raising up would sometimes induce a very light buffet. The spoilers also induced a slight nose-up pitch-up when extended, and this was troublesome during aerial refuelling, where lateral movement to keep station resulted in spoiler deployment, which in turn caused the nose to pitch-up. This combination of desired and unexpected response to a control input was very fatiguing—and frustrating—to the refuelling pilot, and resulted in a later modification.

Flap retraction time was increased slightly, from 40 seconds to one minute, to permit use of smaller, lighter weight electrical drive motors.

There were numerous changes in the cockpit layout as well, reflecting the new panels required by elimination of the ten pneumatic driven hydraulic packs and their replacement by six engine driven hydraulic pumps.

Considerable thought had been given to crew comfort. In the previous B-52s, pilots had traditionally roasted while the navigators froze, leading to a lot of acrimonious intercom exchanges over the position of the cabin temperature control setting. The navigators were provided extra heat, and even the seats were redesigned to lessen the bottom-numbing fatigue of a 20 hour mission. New hot cups for making soup or coffee, as well as new water outlets and relief tubes were provided.

The crew comfort came in handy in a demonstration of the G's capability in Operation Long Jump, by the 5th Bombardment Wing at Travis Air Force Base, California. Commanded by Colonel T. R. Grissom, the B-52G took off at 0704 PST on December 13, 1960, and flew 10,000 miles without refuelling in 19 hours and 47 minutes.

Wichita delivered 193 G models between November 1, 1958 and February 7, 1961, but even as deliveries were going on, the Boeing plant was gearing up for the last and finest series, the B-52H.

Top: *A B-52G in a flight condition that sometimes proved to be critical, down low in mountainous terrain, where wind shear can be catastrophic. (Boeing)*

Above: *The brilliant Pratt & Whitney TF33-P-3 turbofan engine which was to give the B-52H such a tremendous leap in performance. It provided 30% more power, and reduced Specific Fuel Consumption (SFC) from .8 in the J57 to .56. The TF33-P3 was one of those rare instances when you seem to get advantages at both ends of the scale. (United Tech.)*

Above: The shape of the H model changed less perceptibly than that of the G; the nose contour was altered, the engine nacelles had a wide "cowling" like appearance on the front, and the tail gun position now housed a single six barrel 20-mm cannon. (U.S. Air Force)

Right: The "Gatling Gun" can fire 4,000 rounds per minute. (U.S. Air Force)

As remarkable as the J57 engines had been, Pratt & Whitney Aircraft now held out promise of a revolutionary new turbofan (fanjet) engine. The turbofan is one of the very few instances in engineering when one seems to obtain a great advantage without sacrificing something else. By adding a large diameter fan element to the engine (in effect, a propeller, pushing masses of cold air), the total thrust is substantially increased. Specific Fuel Consumption (SFC) in the Pratt & Whitney TF33-P-3 was reduced to ·56 from the ·8 of the J57.

The TF33 was flat rated at 17,000 pounds of thrust, providing the B-52H with a 30% increase in power over the G model using water injection. The thrust rating is calculated at "take-off rated thrust" (TRT) values and with the turbofans, the engine does not exceed 17,000 pounds unless outside air temperature is well above 100 degrees F.

The very quickness of the throttle response posed an unusual and insidious problem to the H model. Too rapid movement of the throttle in certain situations could cause the aircraft to pitch up at a rate beyond the pilot's ability to control with available elevator authority. This is caused in part by fuel slosh—in the B-52H's wet wing, fuel flows slowly back during acceleration, changing the centre of gravity subtly, so that the pilot does not trim sufficiently. After a certain combination of acceleration and "slosh", the aircraft can pitch up, with the possibility of a loss of control.

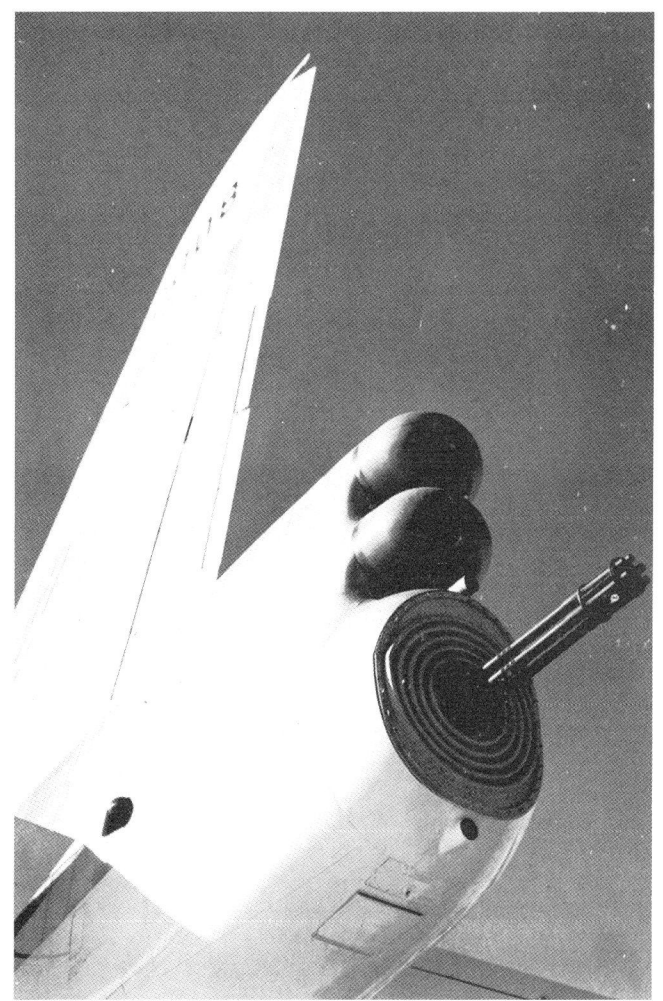

To prevent this, a mechanical thrust gate was placed on the throttle quadrant, which can be positioned for any maximum thrust desired. Once set, the gate prevents the pilot from inadvertently jamming the throttles forward and getting more thrust than he desires.

The air refuelling capability of the H model was also improved by a new spoiler position. The outboard spoiler segments were set to extend up about ten degrees, making small lateral corrections possible without inducing pitch-up.

The B-52H could be distinguished externally from the G model by the very much altered engine nacelles, and by the change from the traditional four machine-gun tail defence to a six barrel 20 mm Gatling gun.

The Douglas Skybolt air launched ballistic missile intended for the B-52H was cancelled after a series of murky political decisions, as a later chapter examines.

The increased range of the new model invited an assault on the non-refuelled distance record held since 1946 by the US Navy Lockheed PV-2 Neptune, the "Truculent Turtle", which had flown 11,235 miles in 55 hours and 17 minutes. On January 11, 1962 a B-52H set a new record by flying 12,532 miles from Kadena Air Force Base, Okinawa, to Torrejon AFB, Madrid, Spain, at an average speed of 575 miles per hour. The flight was called "Operation Persian Rug" and was completed in 22 hours and 9 minutes. The aircraft belonged to the 4136th Strategic Wing at Minot AFB, North Dakota, and was commanded by Major Clyde P. Evely.

The new role of the B-52 as a low level penetrator called forth new equipment and instrumentation for terrain following. The new systems provided relief from the tremendous strain posed on pilots and navigators by flying at low altitudes with inadequate equipment. They permitted the aircraft to be flown at high speeds without the danger of artificially induced structural stresses from too rapid control inputs. The H model was the first to receive the new equipment, although later B-52Ds, Es, Fs and Gs were retrofitted. Later, as we shall see in Chapter Seven, even more modern equipment was installed.

The first generation included the advanced capability radar (ACR) for terrain avoidance, an anti-jamming unit and improved low level mapping capability. ACR gave three dimensional information to a dual mode pilot's display, which were essentially five inch television tubes on which were represented the aircraft's range, elevation and azimuth. The navigator had a similar display. The height of terrain was shown continuously at selected distances of three, six or ten miles from the aircraft.

The pilot could select either a PLAN mode, which depicted a map-like display, or a PROFILE mode, which showed the terrain height at various ranges ahead of the aircraft.

To assist the pilot in what would otherwise be a numbing battle against control forces, control wheel steering was built into the MA-2 autopilot. This reduced the amount of control forces and the frequency of control movements required to fly the aircraft. Instead of using the autopilot trim knobs for control, the pilot continues to use his control wheel and rudders, but with greatly reduced forces. The effect is similar to power steering in an automobile in that some control feel is lost—the controls respond the same at 400 knots as they

Right: In flight the various bumps and bulges of new systems begin to show, along with the antennae of electronic countermeasures gear. (Boeing)

Centre right and bottom right: This is the first B-52H at 410th Bomb Wing, K.I. Sawyer AFB, Michigan, to receive the Electro Optical Viewing System which was to add such a welcome capability to the fleet. EVS package was a long time in coming. It provides both a low light level television sensor and a forward looking infrared sensor (FLIR) so that the pilots can operate with thermal blast shield curtains closed. System also augments the earlier terrain avoidance system. (Jay Miller)

do at 200 knots—but there is a g limiter to prevent over-stressing the aircraft.

It takes courage to bring a 450,000 pound aircraft down to close proximity to the ground, flying at 300 knots in mountainous terrain, under night and instrument conditions. The crews must do a great deal of planning, and there must be instinctive, continuous crew coordination. The pilot and radar navigator must work together to ensure that all instructions are both understood and correctly responded to. Unlike the normal cruise flight, the pilot takes course corrections from the radar operator, and not the navigator, who must concur with each course change. In training, the copilot monitors the instruments, and insofar as possible, maintains a visual lookout.

As hazardous as the operation sounds, the careful pre-planning and crew coordination combine with the equipment to make it a fairly safe procedure, and as it is perhaps the only means of penetrating a modern air defence, the crews recognize the technique's importance.

The low level work also places the aircraft down where clear air turbulence, particularly in mountainous areas, can have its greatest effect. The B-52 had an early history of catastrophic failures due to encountering turbulent conditions. The first of these was when a Boeing crew in a B-52D (56-591) crashed at Burns, Oregon, in 1959, when the horizontal stabilizer failed in low level flight.

Boeing and the Air Force launched immediate programmes to combat the problem. The principal solutions were to avoid all predicted turbulence when possible, and if encountered, to slow the aircraft down and fly attitude control, that is try to keep the aircraft straight and level without making rapid corrections to vertical motion due to gust conditions. All control applications are to be made smoothly and deliberately.

The problem persisted, as will be seen in Chapter Seven, and there were more losses. This, however, was the background against which the B-52H had its introduction, and there would be several additional modifications to the new model.

The 102nd and final B-52H—and incidentally the last production heavy bomber delivered to the United States Air Force— was 61-040, and it was delivered to the 4136th Strategic Wing at Minot AFB on October 26, 1962.

The B-52Gs and Hs had scarcely entered service when they began a long series of structural and equipment updatings that continue to this day. The following table indicates how the installation of additional external equipment has taken its toll on the aircraft in terms of its L/D (Lift to Drag) ratio.

Decline of B-52 L/D

Model	L/D
B-52A/F	21·0 to 1
B-52G/H (clean)	19·0 to 1
B-52G/H (12 SRAM, EVS)*	17·0 to 1
B-52G/H (12 ALCM, EVS)#	17·4 to 1 (improved pylon)

* SRAM—Short Range Attack Missile; EVS—Electro Optical Viewing System
# ALCM—Air Launched Cruise Missile

The increase in drag has been more than compensated for by the increase in penetration and weapons delivery capability.

81

Above left and left: *B-52 cockpit with earlier ACR (advance capability radar) presentation.* (Air Force)

Left: *The improved EVS installation offers a much better picture, and includes much flight instrument information, just as a Head-up Display does.* (U.S. Air Force)

Right: *Crew makes dash for B-52G during Global Shield 79, a comprehensive worldwide SAC bombing exercise. This is at Seymour Johnson Air Force Base, North Carolina.* (U.S. Air Force)

Below: *A B-52H armed with SRAM missiles from 310 Bomb Wing, K.I. Sawyer AFB, Michigan.* (U.S. Air Force)

83

Above: *Occasionally even headquarters unbends and allows a little deviation from standard. This B-52H from the 379th Bomb Wing, Wurtsmith AFB, Michigan, received permission to paint a "triangle K" on its tail during the 1980 Bombing and Navigation Competition. The 379th had used the insignia when flying B-17s from RAF Kimbolton during World War II. (U.S. Air Force)*

Left: *The EVS, the ECM, and especially the missiles have all added up to an increase in drag for the B-52H. (Jay Miller)*

Top right: *The aft gear can be lowered independently, if required. (Jay Miller).*

Centre right: *Side-view of the tail "stinger".*

Right: *Podded engines have served aircraft well since original B-47 conception. They reduce span loading, keep fires isolated from wing and struts act as fences.*

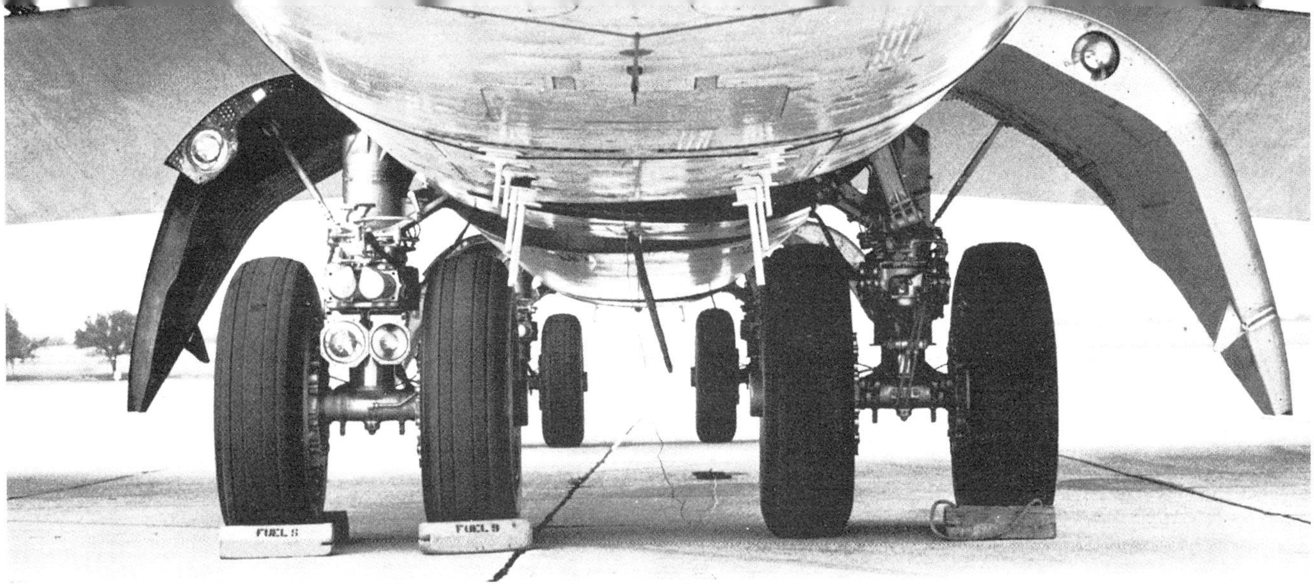
FUELS
FUELS

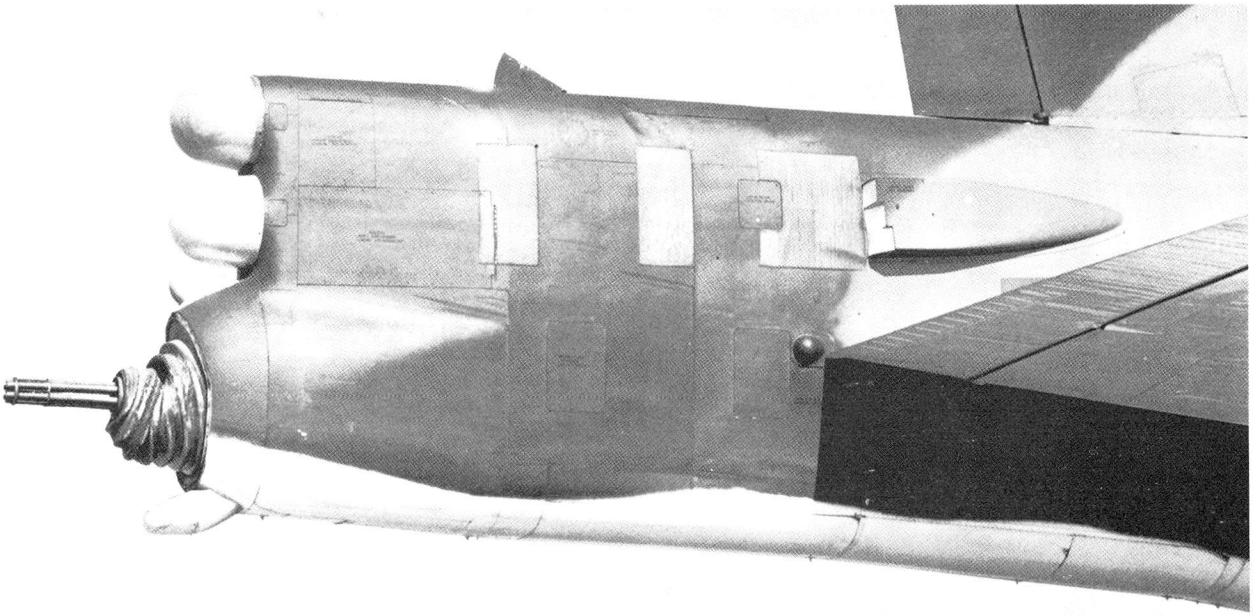

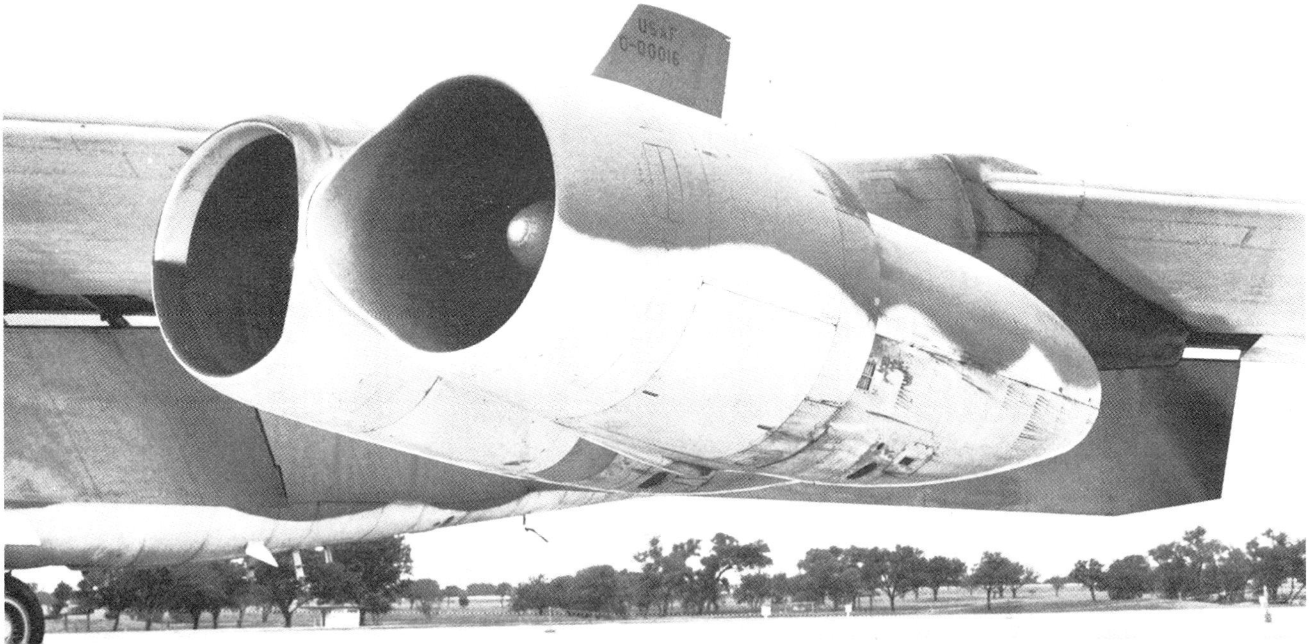
USAF
0-00016

Schematic shows general arrangement of cockpit, B-52G

**CREW COMPARTMENT
UPPER DECK**

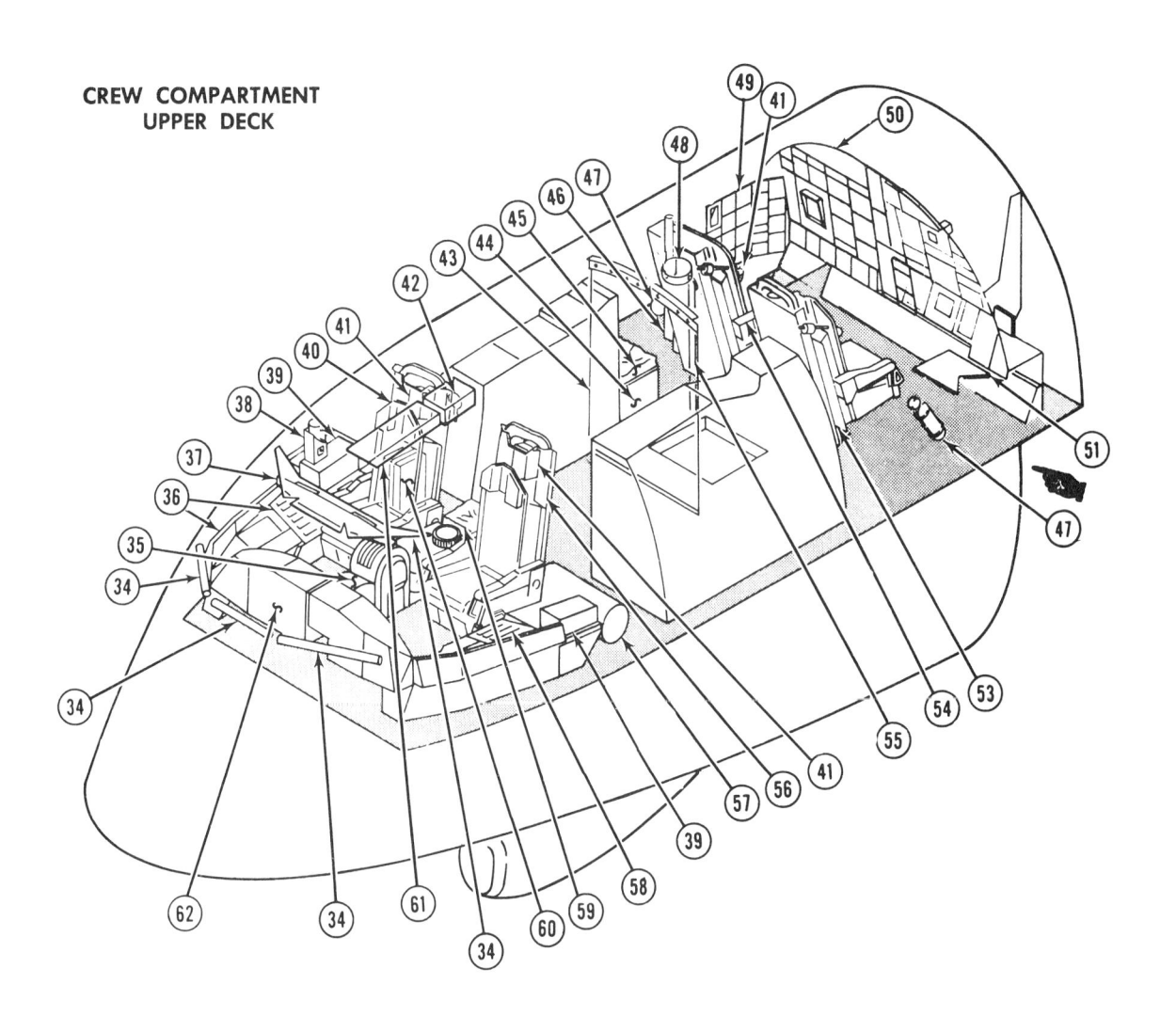

34. THERMAL CURTAIN
35. AISLE STAND
36. COPILOT'S SIDE PANEL
37. EYEBROW INSTRUMENT PANEL
38. HOT CUP
39. FOOD AND DATA BOX
40. COPILOT'S SEAT
41. STATION URINAL
42. SIGNAL LIGHT
43. NIGHT FLYING CURTAIN
44. TOILET
45. DEFENSE INSTRUCTOR'S SEAT
46. FOOD STOWAGE BOX
47. OXYGEN BOTTLE
48. PERISCOPIC SEXTANT MOUNT
49. EW OFFICER'S SIDE PANEL
50. DEFENSE STATION INSTRUMENT PANEL
51. GUNNER'S PULLOUT TABLE
52. (DELETED)
53. GUNNER'S SEAT
54. EW OFFICER'S SEAT
55. STANCHION
56. PILOT'S SEAT
57. MATTRESS STOWAGE
58. PILOT'S SIDE PANEL
59. INSTRUCTOR PILOT'S SEAT
60. PERISCOPIC SEXTANT CARRYING CASE
61. PILOTS' OVERHEAD PANEL
62. PILOTS' INSTRUMENT PANEL

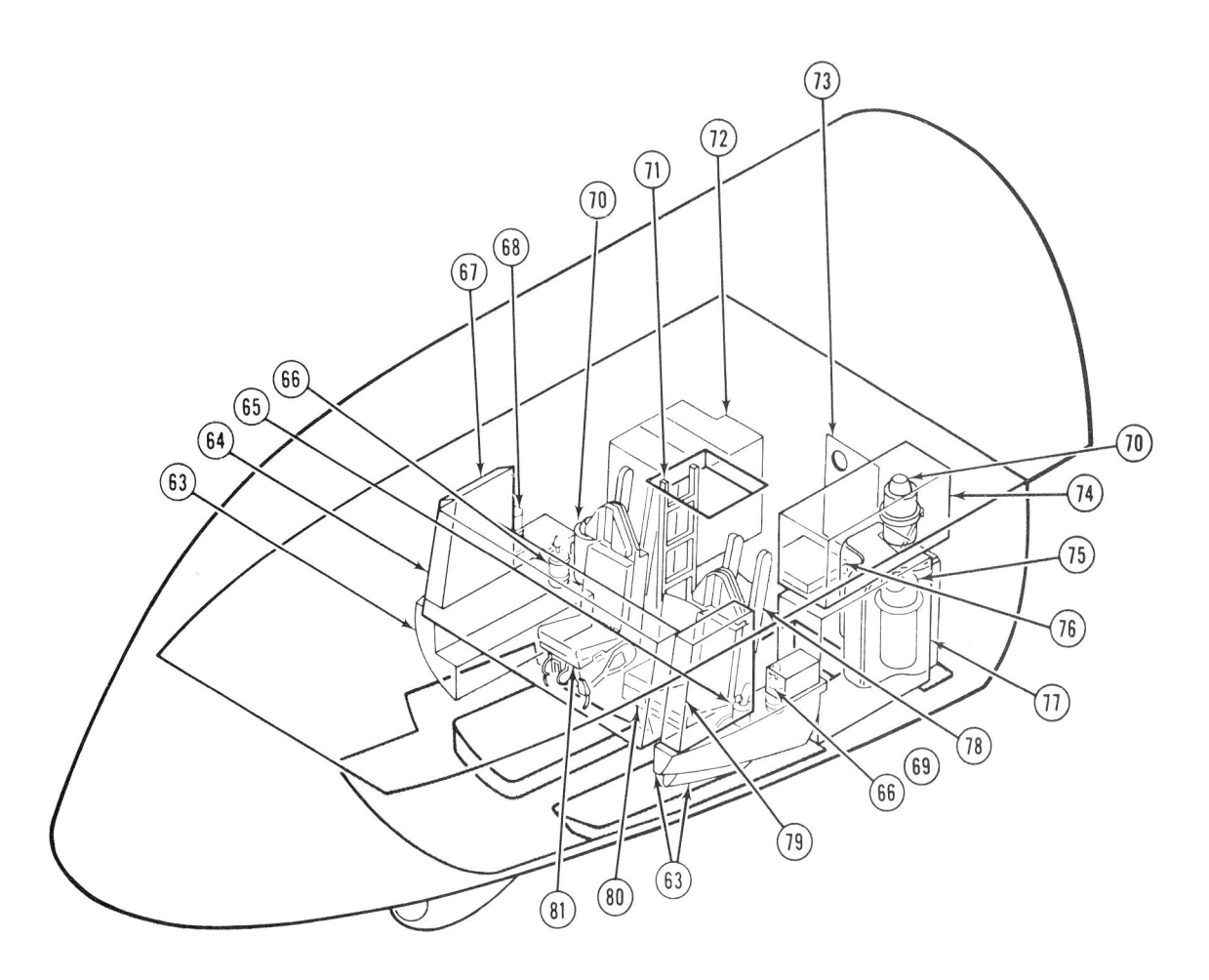

**CREW COMPARTMENT
LOWER DECK**

63. MISCELLANEOUS EQUIPMENT SHELF
64. NAVIGATORS' INSTRUMENT PANEL
65. STATION URINAL
66. OXYGEN BOTTLE
67. NAVIGATOR'S SIDE PANEL
68. HOT CUP
69. FOOD STOWAGE BOX
70. DRINKING WATER CONTAINER
71. LADDER
72. REMOTE MODULES RACK
73. PRESSURE BULKHEAD DOOR
74. ELECTRONIC EQUIPMENT RACK
75. CENTRAL URINAL
76. INSTRUCTOR NAVIGATOR'S TAKEOFF-LANDING SEAT
77. POWER SUPPLY RACK
78. RADAR NAVIGATOR'S SEAT
79. RADAR NAVIGATOR'S SIDE PANEL
80. INSTRUCTOR NAVIGATOR'S DUTY SEAT
81. NAVIGATOR'S SEAT

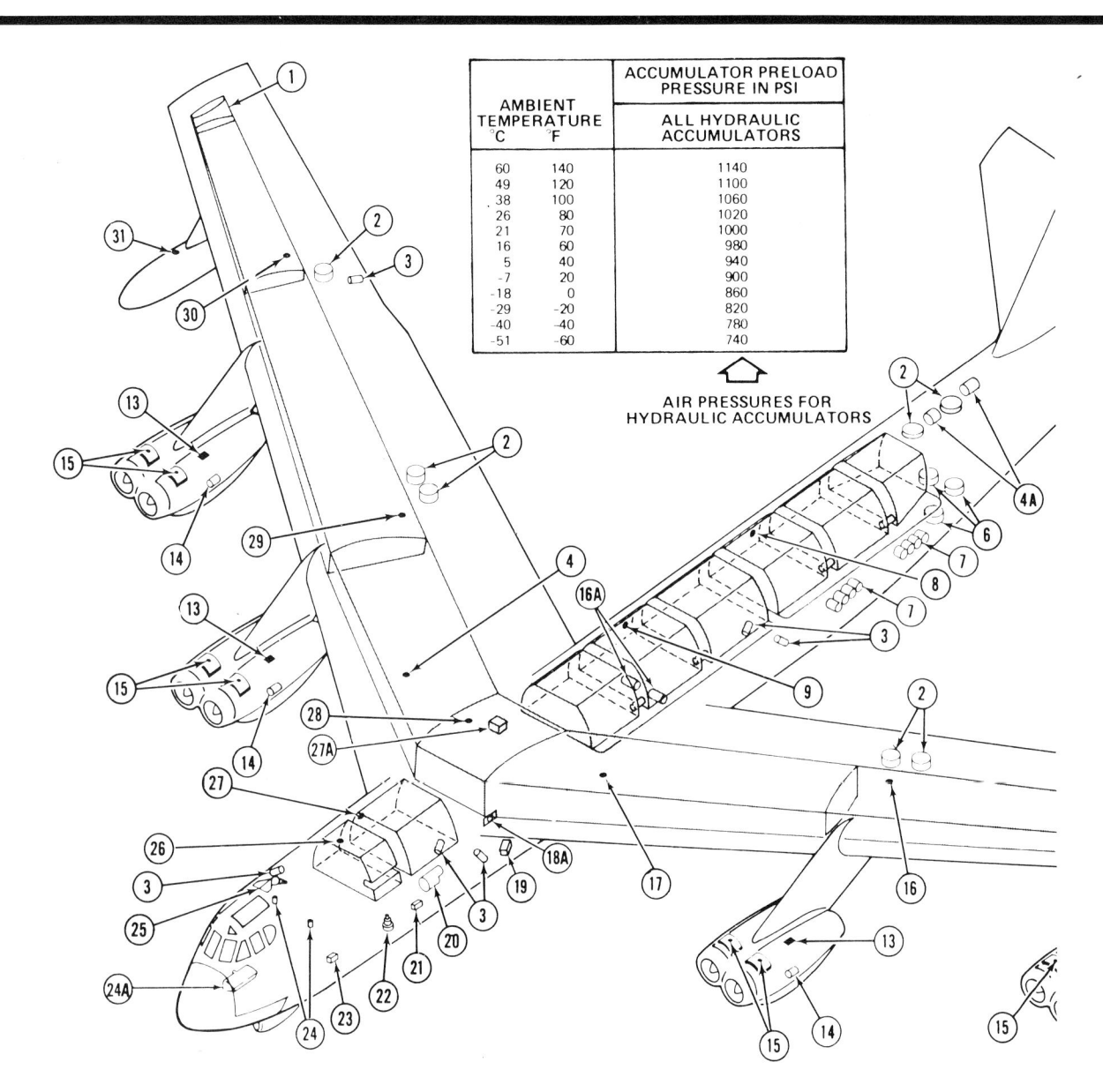

AMBIENT TEMPERATURE		ACCUMULATOR PRELOAD PRESSURE IN PSI
°C	°F	ALL HYDRAULIC ACCUMULATORS
60	140	1140
49	120	1100
38	100	1060
26	80	1020
21	70	1000
16	60	980
5	40	940
-7	20	900
-18	0	860
-29	-20	820
-40	-40	780
-51	-60	740

AIR PRESSURES FOR HYDRAULIC ACCUMULATORS

1. RIGHT WING SURGE TANK
2. HYDRAULIC RESERVOIR
3. ACCUMULATOR
4. NO. 3 MAIN TANK
4A. RUDDER/ELEVATOR HYDRAULIC PUMPS
4B. POWERED RUDDER ACTUATOR
5. DRAG CHUTE
5A. POWERED ELEVATOR ACTUATOR (TYPICAL)
6. LIQUID OXYGEN CONVERTERS
7. STARTER CARTRIDGES
8. AFT BODY TANK
9. MID BODY TANK
10. LEFT WING SURGE TANK
11. LEFT OUTBOARD WING TANK
12. LEFT EXTERNAL TANK
13. AIR BLEED SYSTEM GROUND CONNECTION
14. A-C GENERATOR DRIVE UNIT RESERVOIR
15. ENGINE OIL TANK
16. NO. 1 MAIN TANK
16A. AGM-69A LAUNCHER HYDRAULIC ACCUMULATORS (AIR)
17. NO. 2 MAIN TANK
18. (Deleted)
18A. AGM-69A ENVIRONMENTAL SYSTEM PRESSURIZATION CONNECTOR (AIR)
19. AGM-69A BATTERY
20. AIR CONDITIONING PACK RESERVOIR
21. AFT BATTERY
22. SPR RECEPTACLE
23. FORWARD BATTERY
24. DRINKING WATER CONTAINERS
24A. EVS WINDOW WASH WATER TANK
25. AIR REFUEL RECEPTACLE
26. WATER INJECTION TANK
27. FORWARD BODY TANK
27A. AGM-69A ENVIRONMENTAL CONTROL UNIT (FREON)
28. CENTER WING TANK
29. NO. 4 MAIN TANK
30. RIGHT OUTBOARD WING TANK
31. RIGHT EXTERNAL TANK

Typical servicing of B-52G calls for many ingredients, much equipment, and a great deal of knowledge. It's far more than check the oil and water, please.

THE STRATEGIC AIR COMMAND, FROM BOLLING TO HANOI

Regardless of the remarkable technical features of the B-52, they would only have been a meaningless mass of metal had Strategic Air Command not had the laboriously gathered experience to use their aircraft to their maximum potential, and the imagination and the courage to change their mission as required.

The Strategic Air Command was established on March 21, 1946 as a means of breathing life back into the US Army Air Force bomber commands. The vast aerial armadas of World War II had disappeared like snow in a furnace, and SAC, under the command of General George C. Kenney began operations in that chill spring at Bolling Army Air Base, Washington, DC. SAC strength was limited to 180 Boeing B-29s, 75 North American P-51s and assorted reconnaissance and transport aircraft. Only about 30 of the B-29s could carry atomic weapons, and it is doubtful if there were 30 weapons with which to arm them at the time.

SAC grew rapidly, and it assumed a new and permanent character when Lt General Curtis E. LeMay was named Commander in Chief on October 19, 1948. LeMay would remain in charge until June 30, 1957, the longest tenure of any US military commander, and he would stamp SAC with his personality, his beliefs, his methods and his style. He would change it over from a rather loosely organized pale imitation of World War II practice into the most fearsome military weapon the world has ever known.

LeMay made his person felt at all levels; he demanded the utmost from his people, but he sought to obtain the utmost for them. SAC became an elite force, capable of projecting an awesome degree of power to all the world while undergoing an almost continuous change in equipment and missions. LeMay fostered the development of the B-36, the B-47 and the B-52, and insisted on a degree of crew proficiency and standardization that had never before been attained.

In the next pages the activities of hundreds of thousands of people, the service lives of a dozen different aircraft and missiles, and the products of hundreds of companies are compressed into a sparse recital of the birth, growth, expansion, reluctant decline and final test by fire of the most powerful and perhaps most famous force ever assembled. SAC's striking power for years exceeded not only all the other powers on earth, but all of the powers that had ever preceded it.

Notwithstanding this tremendous force, unparalleled, undreamed of, unimaginable if used, SAC steadfastly remained out of politics, never became a power centre of itself and was always simply the strong right arm of the US Air Force, submissive to the military chain of command and to civilian control.

When the Soviet Union first began to equal and then to exceed its power, when new weapon system after weapon system was proposed and then cancelled, SAC responded by making more demands upon itself and on its people. It trained harder, worked harder, fought harder; it honed its existing weapons and modified them to last yet a few years more. It achieved a record of dignity and honour, and 26 years after it was founded, in the dark days over Hanoi in 1972, it proved its worth in combat.

The demands SAC routinely made upon its people had for years a curiously tonic effect. The postwar pay scales were low, and government housing and other supposed benefits were often non-existent. Aircrews routinely worked 60 to 70 hours per week, and ground crews worked longer. Yet morale in SAC was high, and was reflected in the ever increasing number of hours flown, improved bombing scores, and vastly improved safety records.

LeMay's formidable presence was not diluted by expansion. In 1946, SAC had a little over 37,000 people and 279 tactical aircraft. In 1950, it had 85,473 people and 962 tactical aircraft; by 1959, it had reached 262,609 people and 3,207 aircraft.

The same period saw the introduction of revolutionary aircraft and intercontinental missiles. The concept of missiles was at first foreign to most SAC veterans, but they adapted easily if not eagerly, and the Thor, Atlas, Titan and Minuteman systems were incorporated just as smoothly as the aircraft had been.

At the heart of SAC's success was training, hard, realistic, demanding and constantly evaluated training. The concept of a "joy ride" was unknown. Each mission was planned from take-off to landing to provide maximum proficiency. Each squadron, wing and numbered Air Force had carefully monitored programmes in which every aspect of a combat mission—navigation, aerial refuelling, bombing, gunnery, piloting—was ceaselessly evaluated. Over every commander's head hung the spectre of the SAC Operational Readiness Inspection, when hard boiled professionals would sweep in unannounced and evaluate the unit's response to simulated emergency combat conditions.

During LeMay's tenure the whole pressure system was exhilarating. SAC felt itself to be the first team—it was constantly getting new weapons, it had an expanding budget, and its capability was steadily increasing. The long hours and short pay seemed irrelevant compared to the satisfaction obtained from doing what was obviously an important job.

When General Thomas S. Powers took over from LeMay on July 1, 1957, he knew that it was going to be difficult to maintain the same level of enthusiasm. Training requirements had to become even more rigorous to counter the increase in Soviet capabilities and in 1958 it was necessary to put one-third of SAC forces on ground alert. This meant that one-third of all crews had to be sequestered on the base 24 hours a day for as much as a week at a time, away from their families, totally immersed in the business of being ready to go to war in 15 minutes.

At the same time that alert duty was placing strains on family relationships, SAC was periodically pulsed to a war readiness state by international events. It occurred during the Suez crisis in 1956, and again during the Lebanon crisis in July 1958, when President Eisenhower ordered SAC to a full alert status. It happened again in March 1961, when President Kennedy ordered that 50% of the SAC fleet be maintained on a 15 minute alert basis. SAC had already anticipated a further measure by practicing an airborne alert concept. In 1962, during the Cuban missile crisis in October, the B-52 fleet went into an airborne alert status, flying 24 hour missions, and maintaining a constant airborne striking posture.

In the same year, SAC saw the activation of the first Minuteman missile wing, an event which presaged the dual missile/aircraft nature of the Command in the future. The crews manning the missile sites were to be subject to the same intensity of training, sense of responsibility and fatiguing duty periods that the aircrews experienced, but without the release and exhilaration that flying brought. The missile sites were usually located in remote areas of the northern Mid-West, and it took every ounce of ingenuity the Air Force had to bolster morale.

General John D. Ryan assumed command of SAC on December 1, 1964, a year which saw the B-47 and Atlas missiles start to phase out. With the Minuteman assuming a larger role, tactical aircraft strength dropped from 2,075 in 1964 to 1,490 in 1965.

And in 1965, SAC began a long nine year effort in Southeast Asia. It began innocuously enough with the activation of the 4252nd Strategic Wing at Kadena Air Base, Okinawa, to provide KC-135A tanker refuelling support for Pacific Air Force fighter bombers. This was the beginning of the eventual involvement of the B-52, which would bring about entirely new missions in an entirely new environment, one that would ultimately vindicate the theory of strategic bombing in a savage eleven day campaign.

Simultaneously with the build up in Southeast Asia, SAC had to devise increasingly sophisticated tactics to ensure that the B-52 would survive if an attack were required upon the Soviet Union. Soviet radar, missile and anti-aircraft gun systems had grown at an unprecedented rate, and the high altitudes which had once been the domain of the jet bomber were now forbidden territory. The only hope was to fly low, hugging the terrain to avoid the radar.

There was a third factor which challenged the whole concept of SAC training and orientation. The crew members entering B-52 training at Castle Air Force Base were no longer veterans of several thousand hours of flying experience; many were fresh from flying schools, and they lacked the rigorous training of their predecessors. The demand on instructors and commanders increased in direct proportion to the level of inexperience. Retention of veteran personnel was at once more critical and more difficult, for a pilot could earn at least three times the pay for flying for the airlines, while the airman serving as an electronic technician could obtain four times his salary on the outside.

There was a fourth element, also. Inflation and a levelling off of appropriations had greatly reduced SAC's budget, especially in the operations and maintenance areas.

The SAC response to this four fold challenge was typical: take more out of its hide, from the substance of the force, the people. More effort, more hours, more training and more sacrifice were the only answers available. In the long run the

Right: *The ultimate application of B-52 combat capability turned out to be totally unforeseen; no-one could have imagined that their combat debut would be dropping iron bombs in a tiny Asian country that many Americans had never heard of.* (Randall D. Thompson)

almost draconian training standards paid off in the high level of standardized crew ability that had never been achieved before on such an enormous scale. It would permit crew members from a half a dozen bases around the United States to be thrown together in Southeast Asia, able to operate as a unit from the start.

There was another, unpredictable effect, one that perplexed even the crew members themselves. This was the intense loyalty to the crew as an operating unit, a loyalty that transcended individual personal likes or dislikes. In Southeast Asia, the crews truly lived together, played together and flew together. In a violation of a universal military custom, the gunner, an enlisted man, was quartered with the otherwise all-officer crew, and was a part of the social unit as well.

Below: *There were many anomalies besides the fact that a high altitude nuclear bomber was dropping 500 and 750 pound iron bombs. The most amazing of all was that most of the tonnage was dropped inside the territory of the ally we were defending, South Viet Nam. The pocked mark path of a B-52 raid in 1965, 20 miles northwest of Bien Hoa in 1965 was the foretaste of a seven year frustration. The invading North Vietnamese and their Viet Cong guerilla allies grew to fear the B-52 above all weapons.* (U.S. Air Force)

In combat, the crews naturally drew closer together. The tours were long—some crews spent as many as 1,100 days at either U Tapao, or Guam, or both, in a six year period.

The crew identity made it possible for them to withstand combat conditions that might otherwise have caused severe morale problems. The closeness of the group was so great that on the semi-annual leaves, after the first day or so of reunion with the families, the crews would suddenly find themselves together again at the club, or at the pool, drawn by the comradeship that had grown up. It puzzled the wives and children at first, but they soon came to understand and accept it. The importance of the crew concept will be alluded to again, below.

The first B-52 engagement in Southeast Asia was not auspicious. On June 18th, 1965, 27 B-52Fs from Carswell Air Force Base's 7th Bomb Wing and Mather Air Force Base's 320th Bomb Wing made the initial BUFF attack. Operating from Andersen Air Force Base, Guam, the units dropped conventional 750 pound and 1,000 pound "iron bombs" against an unseen Viet Cong target.

Despite the usual complimentary reports from ground commanders, it was not evident that the raids had been a great success, and there was much natural scepticism about the utility of a high level radar bomb drop against guerilla troops. Further, two aircraft had been lost in a mid-air collision.

Yet within a few months there was a universal acceptance

Above: *The supply train from the U.S. munition factories to the Vietnamese jungles was enormous, involving thousands of people and millions of dollars. There were innumerable problems, from letting contracts, to building the bombs to securing the shipping to moving them by rail and truck from the ports, to storage at the bases, to build up at the Armament Squadrons, to loading on the plane. The B-52s were able to off-load the bombs with such clock-like regularity that it was difficult to build up an inventory of more than a few day's supplies. Thus every shipment, every rail car of bombs, became a critical item to monitor on the management charts. Bomb assembly areas like this one on Guam worked around the clock to keep the right mixture of munitions available.* (U.S. Air Force, via Brig. Gen. J. R. McCarthy)

Centre right: *The bombs could be transported on flatbeds like this for individual loading.* (U.S. Air Force, via Lt. Col. George Allison)

Right: *A lot of muscle was required despite all the modern handling gear. These are 750 pound bombs. The bomb loader was referred to as a "jammer" because it "jammed" the bomb in the bomb bay.* (U.S. Air Force via Brig. Gen. J. R. McCarthy)

of the B-52 as a new and potent artillery of the air, and by November 1965, the B-52s were able to support directly the 1st Air Cavalry Division in mopping up operations near Pleiku. The B-52s were employed in what were termed "Arc Light" operations, and unleashed 1,796 tons of bombs on the enemy, the equivalent of a World War I artillery barrage.

As we shall see in the next chapter, SAC had not been fully prepared to fight an "iron bomb war", and a programme coded "Big Belly" was underway to modify B-52Ds to carry substantially more bombs. With the introduction of the Ds a new ground directed bombing system called "Combat Skyspot" provided tremendously increased flexibility in targeting, and vastly improved accuracy over the airborne radar systems.

One year after commencing operations, B-52s were dropping an average of 8,000 tons of bombs per month on Viet Cong targets. Results were often difficult to verify, and astute Naval commanders sometimes inquired as to what was hit instead of how many tons were dropped, but intelligence sources reported that the B-52 was regarded as the most feared weapon available to South Vietnamese forces.

By 1967, when General Joseph Nazarro assumed command of SAC, B-52s were flying even more sorties, and had dropped more than 190,000 tons of bombs in two years of operation. U Tapao Royal Thai Navy Airbase, Thailand, became available as a SAC base, and sorties could be flown in a little under four hours, compared to the 16 to 18 hours required from Guam.

A Flying Tiger veteran, General Bruce K. Holloway, replaced General Nazarro in 1968 as CINCSAC and directed the spreading out of Southeast Asia B-52 crew duties by creating a Replacement Training Unit at Castle Air Force

Above: *Bombs could also be preloaded in "clip-in" installations at the ammunition assembly areas.* (U.S. Air Force via Lt. Col. George Allison)

Left: *"Project Big Belly" had resulted in an extremely compact arrangement within the B-52D fuselage which permitted as many as eighty-four 500 pound bombs to be carried internally, with another 12 on each of the two pylon mounts. Not only space was critical, but also the release sequence, so that the bombs would not bump each other as they dropped.* (U.S. Air Force via Brig. Gen. J. R. McCarthy)

Base, California. SAC crews, who previously might have flown the advanced G or H models, were sent through an intensive two week course on B-52Ds, making them eligible for duty in Southeast Asia. There began a systematic rotation of crews among parent units to the Arc Light operations, which had assumed an unreal permanence.

1968 saw the B-52s flying missions in the relief of Khe San, where unbroken waves of six aircraft, attacking every three hours, dropped bombs as close as 900 feet to friendly lines. A combination of rigid scheduling and even greater output on the part of crews allowed an increase to 1,800 B-52 sorties per month.

Day after day, B-52s would take off from U Tapao, dropping bombs on supply dumps, base areas and troop concentrations. The B-52 war spread to Laos and Cambodia in pursuit of the North Vietnamese infiltration network. From being a precision "long rifle", the B-52 had become the workhorse of the war.

The crew member's life during this period was one of alternate fatigue, tension and boredom. The hours were long, the mission exacting, and the almost perpetual six month separations from home were painful. It seemed almost as if they had stepped out of life in the real Air Force and begun an independent, never ending cycle of brief, fly, sleep and brief, fly sleep. The ground crews worked a 72 hour week routinely, and worked 84 hours during surge periods, struggling with the monotonous but deadly routine of

Below: *The massive "Arc Light", "Bullet Shot" and "Linebacker II" operations called for diligent preplanning and careful execution of the flight plan to permit aircraft to assemble in such a manner as to hit their bomb drop times exactly on schedule. Timing was often crucial because of other aircraft scheduled to drop, or because the drop time was keyed to some allied ground operation. On the flights from Guam, the differences in wind and weather required the navigators to be absolutely on the mark, able to devise either extensions or short cuts to the routing which would place them exactly at the right place at the right time. Typical Arc Light mission shown here. (U.S. Air Force via Lt. Col. George Allison)*

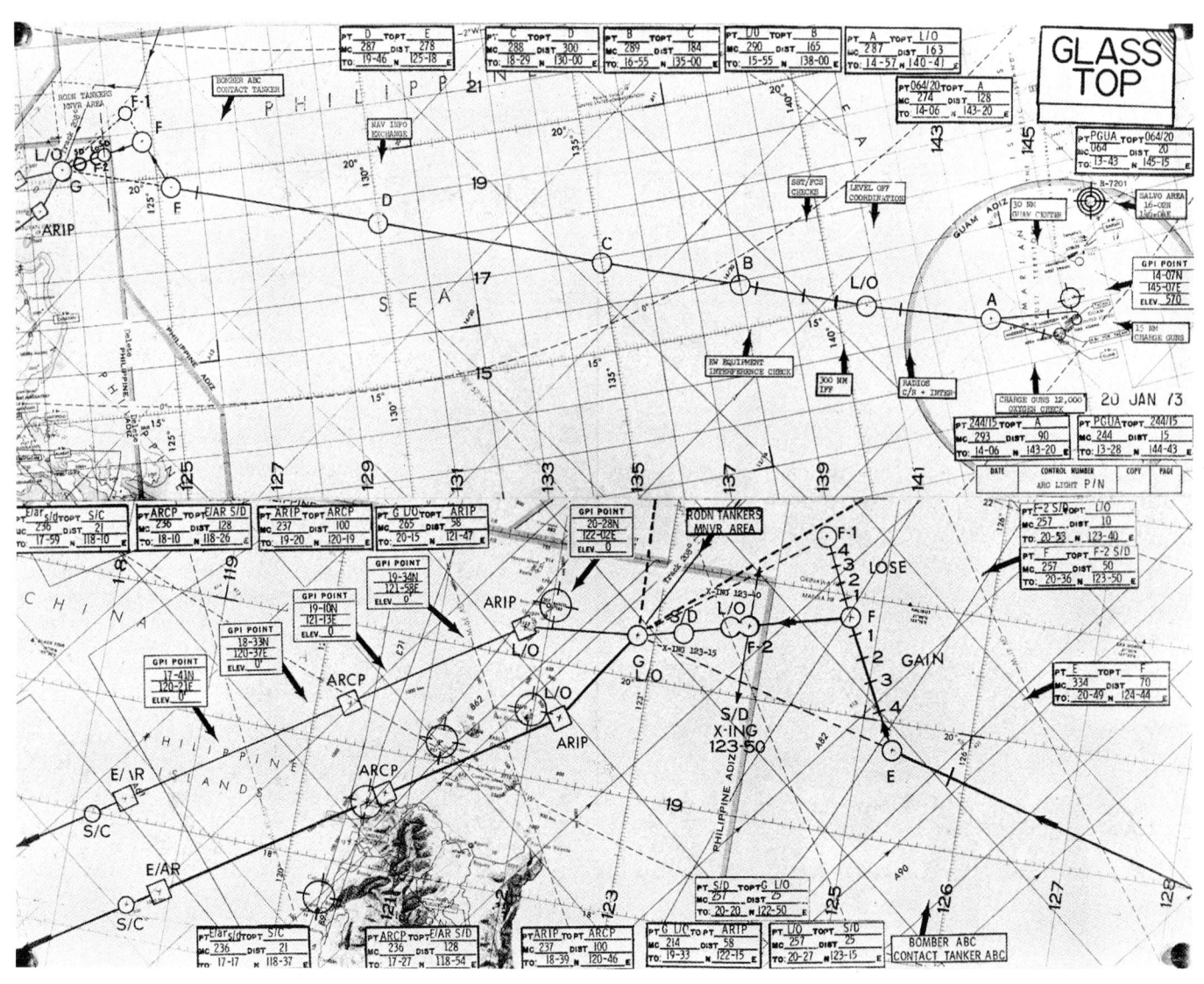

94

assembling bombs, maintaining the aircraft, and waiting for the 179 days to roll around. (The 179 day period was an administrative solution to the legal problems involved. If an airman or officer was sent overseas for more than 180 days, it had to be considered a "permanent change of station" which was expensive and disruptive.)

While the Southeast Asia effort went on, SAC's nuclear deterrent was undiminished, with FB-111s being phased in and B-58s being phased out.

World War II ace General John C. Meyer took over command of SAC on May 1, 1972. General Meyer would see his forces effectively bring an end to the B-52 war in Southeast Asia, and on a note no-one would have predicted: the absolute mauling of the North Vietnamese air defence, and the precision bombing of Hanoi, Haiphong and other targets during the B-52's finest hour—"Operation Linebacker II".

"Linebacker II" came into being reluctantly, as the last military option but one—the nuclear option—that the United States had to disengage itself from the increasingly unpopular war in Vietnam. Henry Kissinger, Assistant to the President for National Security, had been involved in negotiations with the North Vietnamese leader, Le Duc Tho, over a period of months, trying to establish a framework by which the United States could leave North Vietnam and South Vietnam nominally at peace, and with some sort of military parity, so that the US could withdraw its forces.

As the months dragged on, it became apparent that North Vietnam was as interested in protracting the discussions as the US was in concluding them. There was a rising current of domestic unrest in the United States that seemed to run to North Vietnam's advantage.

In the crudest terms, the US was faced with a situation in which the enemy literally had to be driven to the negotiating table, to be beaten until the prospect of a US withdrawal and a cease-fire was more attractive than endless guerilla warfare in the South.

The decision was made to employ the B-52 for the first time as it would have been employed all along by the military commanders, had they not been restricted by political rules of engagement. The B-52s were to be put over the heart of the enemy *en masse*, with maximum effort, in a *schwerpunkt* that has been the heart and soul of strategy since war began.

Above: *A relatively rare shot of a B-52 F dropping bombs in South Viet Nam in 1965. When "Big Belly" gave D models much greater capability, Fs were withdrawn.* (U.S. Air Force)

Above right: *Waiting on the ramp in 100 degree weather, ground crew's tasks are temporarily done. The man on the right is a newcomer – in a few weeks he would have a deep tan.* (Courtesy Major Dwight A. Moore)

Right: *A B-52D taxies out. Note how the wings droop with weight of bombs and fuel.* (Courtesy Major Dwight A. Moore)

It could not be done overnight. Even though the long build-up in Southeast Asia had established a mammoth supply chain of ships, aircraft, trucks and railroads, carrying the thousands of tons of bombs, fuel and other supplies to US forces, the full potential of the B-52 could not be applied without a build-up from the United States.

Activity began to grow from February 1972, as the Eighth Air Force, the same Air Force that had destroyed Nazi Germany, commenced "Operation Bullet Shot". Commanded by a leading World War II Thunderbolt ace, Lt General Gerald W. Johnson, "Bullet Shot" would see the resumption of operations from Andersen Air Force Base, Guam.

Operations from Guam were intrinsically more difficult than operations from Thailand. The mission day was 18 or more hours long; a refuelling was involved, and the wear and tear on the aircraft was much greater, creating a higher maintenance workload. As entire wings rotated into Guam—the 7th from Carswell, the 306th from McCoy, the 96th from Dyess—additional sorties were flown. In time every B-52D unit in the US, with their new massive capacity of 108 bombs per aircraft, would be assembled at Guam. With the additional B-52Gs called in, there were 155 aircraft spread over five miles of parking ramp, a powerful assemblage of ageing hardware. Aircraft parking space became so crowded that a standard joke was that at least ten planes had to be kept taxiing at all times, as there was no place to park.

The B-52Gs which supplemented the force were more modern aircraft than the Ds, and had a greater range, but they carried only 27 bombs and in combat proved to have less electronic countermeasure capability.

The pressure on the enemy was reflected by the pressure on the ground crews supporting the continually mounting effort. The 303rd Consolidated Maintenance Wing at Andersen became the largest maintenance unit in SAC, with more than 5,000 personnel. Most of these would spend 179 days on the Rock, as Guam was unaffectionately called, go home for 30 days leave (five or six of which was spent in travel), and then return for another 179 days of unremitting labour.

The stress was similar at U Tapao, and at both locations, unorthodox methods were used to speed up maintenance. The preflight checks which took eight hours in the States were done in four in Southeast Asia.

As experience built up, the combined bomb dropping efforts of U Tapao and Andersen broke the North Vietnamese offensive of April through June, 1972. Sortie rates soared to an unbelievable 3,150 per month.

All of this was "Bullet Shot"; it was but a prelude to the final major effort, "Linebacker II". ("Linebacker I" had been an earlier, but much more limited series of strikes against North Vietnamese targets.)

"Linebacker II"

The original conception had been for three days of maximum effort over Hanoi and Haiphong, with support from other Air and Naval forces. The method of attack was to be night radar bombing from high altitudes of purely military targets. Each raid would consist of three waves of B-52s, of varying strengths, each hitting targets at four to five hour intervals.

Right: *Andersen Air Force Base, Guam. The field was absolutely packed with aircraft, some in revetments, others in unprotected areas.* (U.S. Air Force via Lt. Col. George Allison)

The initial attack called for 42 B-52Ds from U Tapao, and 54 B-52Gs and 33 B-52Ds from Andersen. A total of 93 sorties were planned for the second day, to be followed by 99 on the third.

Despite the tremendous experience built up during "Bullet Shot", "Linebacker II" called for even more ingenuity as the take-off and recovery times were more compressed. The two bases were each to put out more sorties in a single night than a stateside SAC base did in a month.

For the "Linebacker II" missions, a new concept of mission importance defined as "Press On" was adopted. Prior to this, those missions which encountered any mechanical malfunction which would in anyway compromise flying safety would automatically be aborted. Because of the national importance of "Linebacker II" these standards were replaced by "Press On" rules. Maintenance problems which would have caused a delay or an abort in the US were either ignored, or resolved in an unorthodox manner. The criteria for "Press On" decisions were simply these: could the aircraft get over the target, and could it release its bombs? If the answer was yes to these two questions, the decision was to go. Problems of what might happen afterward were of course considered, and every attempt made to find a solution, but even severe emergencies which would have dictated a precautionary landing in ordinary circumstances—the loss of two engines, for example—were not considered enough reason to abort a "Linebacker II" mission.

Ground crews were sometimes kept aboard the aircraft to continue maintenance *en route* to the target; spare crews and spare aircraft were always standing by to take the place of someone forced to drop out. The aircraft took off and flew in three ship cells (a loose, radar-station keeping formation which provided maximum ECM coverage) and if an aircraft aborted and taxied off the runway, a substitute from the following cell would "roll forward" to take its place. Its position in its own cell would be taken by another B-52 from the one behind it. This "roll forward" was the only expeditious way to replace aircraft, for the taxiways and parking areas were far too crowded to do it by conventional means. The resultant change in cell composition meant that the crews had to rebrief themselves on the spot, for the targets and/or target times might be far different than had been assigned. Only the crews with the training and flexibility of SAC would have had the ability to change missions, rebrief, and then conduct the mission with precision.

There were some disquieting tragic aftermaths to the "roll forward" practice. When the process occurred several times, the identity of individual crews and aircraft were lost as they assumed new positions. If, over the target, an aircraft was hit, it was identified by its cell and number— Rose Two, or Amber One, for example—but not by its aircraft number. There might be further confusion if some of the aircraft diverted to U Tapao or to a base in South Vietnam.

As a result, combat casualties were not always known immediately by the families who lived on Guam. Each aircraft lost represented six casualties, six families to be notified. On the morning after a raid, before the crews got back from debriefing, there was an ominous pall over the area where families lived. They would go on with the ordinary process of living, going to school, to the pool, shopping, but always with an eye for either the return of their loved one, or the dispiriting sight of a caravan of staff cars, carrying the Chaplain, the Base Commander or his representative, and most of all, carrying the tragic news of one who was now lost. The spectacle of the cars offended some, who felt it could be handled in a less traditional, less disquieting way. Others took it as it was intended, the official, genuinely caring expression of sympathy and shared loss.

"Charlie Tower"

Another important maintenance/operational innovation which aided the all-out effort was "Charlie Tower", where the Wing Deputy Commander for Operations established an on the spot control of all bomber aircraft on the ramp and in the air for maintenance instructions. (Control of the mission itself was retained at the Command Post.) The function of "Charlie Tower" was to control the movement of the aircraft around the airfield so that the orchestrated take-off sequence of three ship cells could take place, maintenance on the aircraft being groomed for the next mission could be accomplished, and advice on how to handle inflight or ground maintenance problems could be given.

The most experienced experts on each of the aircraft systems were either on hand or on call by telephone, and there was even a satellite patch back to the Boeing experts at Wichita, Kansas. This group, headed by a genial genius named Lawrence D. Lee, could muster the entire expertise of the Boeing company to analyze, in real time, the inflight problems of a B-52 in combat. Lee recalls one incident where he received a call at two o'clock in the morning to provide the best glide speed for a B-52 with all engines out. It was trying to stretch its glide back to friendly territory; four engines were out and fuel leaks would cause the other four to flame out shortly. Lee asked a couple of questions about weight, and then came up with the right answer: use the maximum endurance speed for the weight of the aircraft. The aircraft made it back, all engines out, just over the border of Thailand, where the crew bailed out.

The first pilot to take off on the first "Linebacker II" mission on December 18, 1972, was Major Bill Stocker in Rose I (the three ship cells were coded by colour, from prosaic Red and Blue to the more exotic Cherry, Slate or Ebony). He knew he was leading an armada which would be joined by a host of other US combat elements. There would be Douglas EB-66s, Douglas EA-3s, the bulbous old "Skynights", and Grumman EA-6s to assist in jamming enemy frequencies. Flights of McDonnell Douglas F-4s, General Dynamic F-111s and Vought A-7 aircraft would attack airfields and keep the MiGs down. Republic F-105 "Thuds", very sophisticated Wild Weasel aircraft, would use their Shrike radar homing missiles to suppress SAM sites. The fighter F-4s would be on hand to serve as a MiG combat air patrol, an air superiority force which would also provide escort home for any crippled B-52s.

Further out, tired old EC-121 Connies, their engines perpetually dripping oil, would monitor the frequencies and plot surface to air missile firings. Lockheed C-130s and Jolly Green Giants, the Sikorsky HH-53 choppers, would be on hand to provide rescue operations. As always, the Boeing KC-135A tankers of Kadena's 376th Aerial Refueling Wing would provide pre-strike fuel for the thirsty D models, and if necessary, emergency post-strike refuelling.

Ships of the Seventh Fleet added to the monitoring and rescue capability. One ship, totally unknown to the crews other than by its call sign "Red Crown", provided such extraordinary ground and air surveillance that it is still talked about admiringly in SAC ready rooms.

An ominous complicating factor to the operations, and one which still provokes anger from veteran crew members, were the instructions from SAC Headquarters that the aircraft were *not* to take evasive action either from SAMS or MiGs during the long run in from the Initial Point (IP) to Bombs Away. The concern was two-fold. First, SAC wanted to be sure that only military targets were hit, and second, it was felt that the electronic countermeasures integrity of the three ship cell might be lost if evasive action were taken.

There are sharply divided opinions about these instructions. One side maintains that the basic SAC tactical doctrine called for the specific evasive action manoeuvres which would have optimized the cell's ECM capability, while another side maintains that the manoeuvres which were ultimately used were devised on an *ad hoc* basis in the light of accumulating experience. In any case, losses were to be severe.

When the bombers penetrated the target area, SAMs lit up the sky around Hanoi. A firing could be detected by a white city-block-square sized flash on the ground, followed by a streak through the sky that finished, in the clouds, in a mushroom shaped halo of exploding light of enormous proportions. The SAMs were plotted on the electronic equipment, too, signals from the ground to the missile (Uplink) and from the missile to the ground (Downlink) both being detectable. The crews would sit listening to the radio chatter signalling the launches, and could see out of the windscreen "wall to wall SAMs" as one pilot put it. The Electronic Warfare Officer would pick up on those signals that indicated a missile was closing. Thirty or forty or more SAMs were sometimes in flight all at once, all apparently aimed at one cell. One crew calmly broadcast "this one is going to get us", and then went on to drop its bombs on target before being hit. It was a miracle of iron nerves and discipline.

For some "EWOs", Electronic Warfare Officers, the signals were the first real representations of an electronic phenomenon that they had been practicing under simulated conditions for years; they took a real delight in the clarity of the signal presentation, fascinated by something they had been trained for all their careers suddenly coming to life, and ignoring the implicit danger.

The first attack broke new ground for both sides. Over 200 SAMs were fired, and two B-52Gs from Andersen and a B-52D from U Tapao were lost.

The data from debriefed crews was assessed as soon as possible, but the second day's first wave was launched on the basis of the pre-strike plan. It encountered heavy SAM opposition, but no losses. It was evident that the dictum of no evasive action from IP to Bombs Away was too costly. The word was sent to the second and third waves that coordinated evasive action could be taken, if ECM integrity was maintained, and if the bomber was straight and level immediately prior to bombs away.

Data analysis provided some new methods for jamming the SAMs, and on better use of evasive tactics. New black boxes were conjured up and installed. It was hoped that the third attack would benefit from all the information, but instead it incurred the highest losses of the war. A total of four B-52Gs and two B-52Ds were shot down out of the 93 aircraft launched. The aircraft had made their customary breakaway over the target, and turned into over a 100 knot jet stream that seemed to stop them in their tracks. In addition, the breakaway banked the aircraft to a critical attitude that blanked out some of their ECM capability.

MiG-21s had been observed flying parallel to the bomber formations, just as the Germans had done in World War II, radioing speeds, altitudes and headings back to the SAM sites. At least 220 SAMs were fired, some in salvoes at the turn-point, when the bombers were most vulnerable.

The losses caused tremendous concern at every level of headquarters, and in the national press. General Meyer, CINCSAC, went over every detail of the last three days' efforts, analyzing every aspect, US and North Vietnamese. Six B-52s out of a force of 93 was almost a 7% loss rate. Three or 5% was considered acceptable; more than this was not, for there was no B-52 production line any longer, and the US could not afford to sacrifice either its nuclear deterrent or its heavy artillery to the SAMs of Hanoi.

Meyer decided that the enemy had been hurt, too, and that the 8th Air Force would press on. Instead of the three days originally planned, the intensity of bombing would be sustained for an indefinite period. Ground crews would get no respite as they went through the endless process of refuelling, rearming, and servicing the bombers; aircrews would fly every two or three days. Meyer's decision to go on was the correct one. The new tactics, which had featured destruction of SAM sites, coupled with the depletion of the SAM inventory, had turned the tide decisively for the B-52. The bomber effort of the next two missions was confined, as planned, to U Tapoa's B-52Ds, while Andersen began preparing for another maximum effort.

During this period a decision had been made to seek out and destroy the SAM storage depots, in an effort to cut off supplies at the source. The enemy defences seemed to be

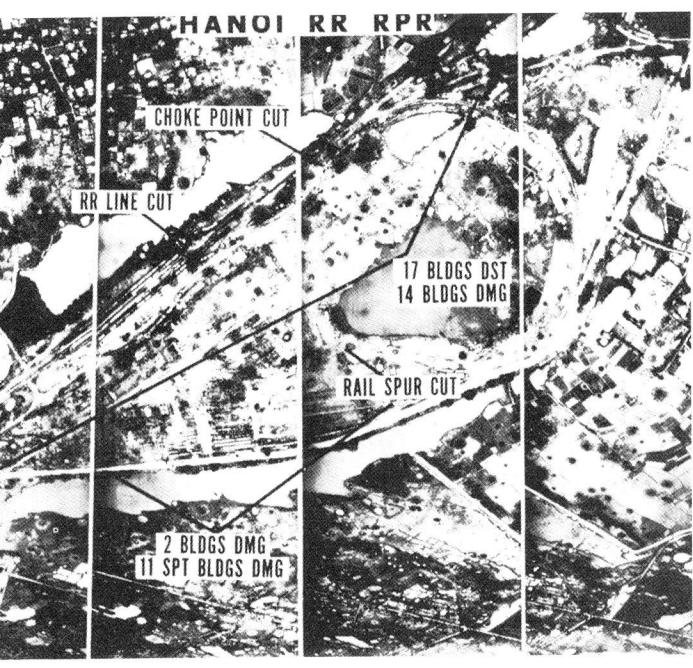

Top: *The long lines of B-52s lined up, tyres squeaking, engines belching JP-4 laden fumes, brakes complaining, crews sweating, radio filled with marshalling instructions, a tight nervous feeling among everyone concerned. This was December 26th, 1972; 120 B-52s would arrive over North Vietnamese targets at the same instant.* (U.S. Air Force via Lt. Col. George Allison).

Above: *Bomb damage assessment photos of the Hanoi railroad yards after December 26th raid.* (Air Force via Lt. Col. George Allison)

Right: *BDA of Hanoi/Gia Lam airfield.* (Air Force via Lt. Col. George Allison)

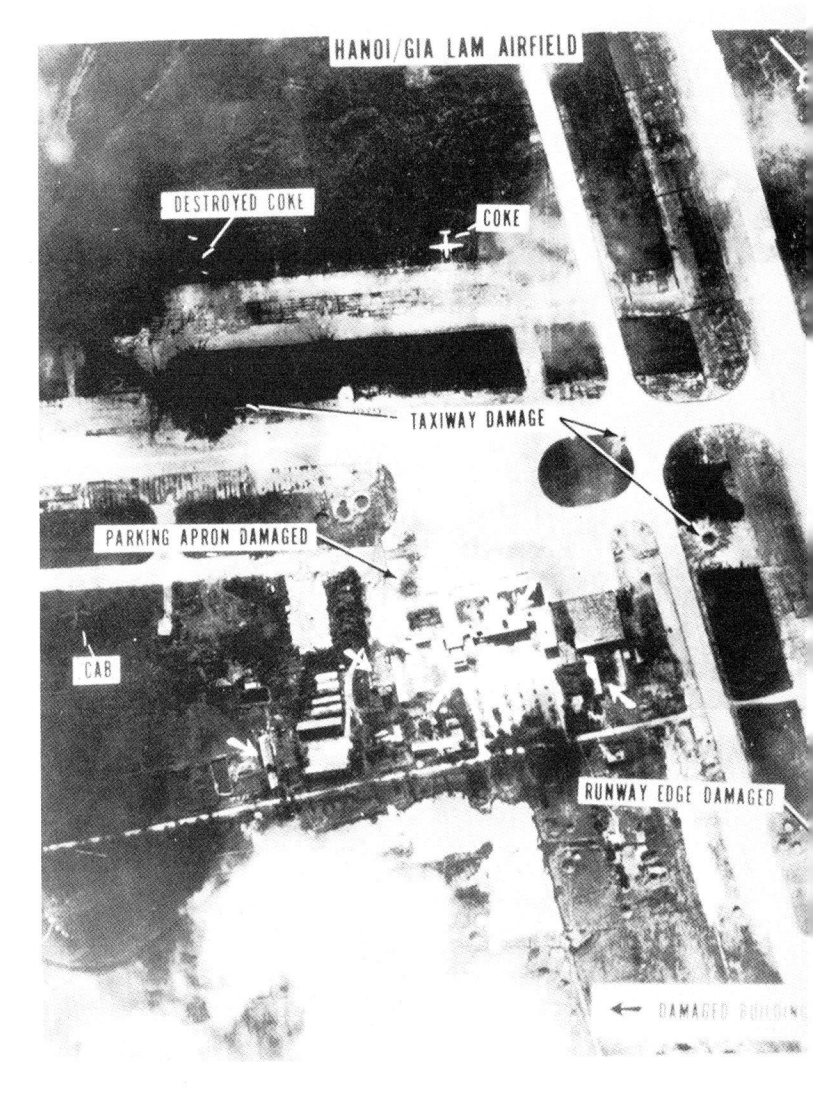

having trouble coping with the much more varied tactics being employed by the BUFFs. Three successive days went by without any B-52 losses or damage from SAMs. The MiG-21s, which had occasioned real worry prior to the start of "Linebacker II", had not shown any aggressive desire to engage the bombers, and at least two had been shot down, one by T/Sgt Sam Turner and another by A1C Albert More. Three others were claimed but not confirmed.

Christmas day saw a stand down; crews at both U Tapao and Andersen had a day's respite, and even the ground crews got a few hours off. To many, retrospectively, it was a bitter mistake, for the North Vietnamese spent Christmas transporting SAMs to the firing sites, catching their breath for the inevitable resumption.

On December 26, entirely new tactics were tried. All Andersen aircraft, 33 B-52Ds and 45 B-52Gs, were launched in a compressed 2 hour and 29 minute period. Four waves of these aircraft, from four different directions, were to converge simultaneously over Hanoi, while three other waves were to strike Haiphong at the same time. There were 120 targets, all with the same bomb release time and there were 110 support aircraft in the air.

Colonel James R. McCarthy, Commander of the 43rd Strategic Wing, was airborne commander for this raid, despite a raging case of pneumonia. The attack was heavily opposed by SAM launches. Weather conditions were difficult, with 100 knot winds which made the simultaneous release of bombs from forces approaching from opposite directions doubly difficult. Ground speeds differed by as much as 200 knots, and the navigators had their hands full to bring their aircraft into position on time.

The attack went off with precision, bombs dropping on target at exactly the predicted times; despite the heavy increase in SAM firings, only two B-52s were lost. The ascendancy of SAC over Hanoi was established.

For the next two days, the B-52s turned eagerly, with the fighters, to the destruction of individual SAM sites, and henceforth the accuracy of the missile firings declined noticeably. The last two B-52s to be lost were shot down on the ninth day of the raids.

The tenth and eleventh raids were without incident; the bombs went on target, primarily SAM storage depots, and the word came down from Headquarters that "Linebacker II" was over. The North Vietnamese had returned to the negotiating table. The US was to be able to achieve its objective of letting go of the war. The B-52s had succeeded where all other tactics had failed.

In those 11 days, the only days that SAC fought the war as it wanted to fight it, in ageing airplanes, some over 17 years old, SAC had flown 729 sorties across the most heavily

Right: A SAM detonates very near an F-4 fighter. Fighters had a chance to evade SAMS through violent manoeuvres. B-52s were restricted to relatively gentle manoeuvres and concentrated ECM emissions. (US Air Force, via Lt. Col. George Allison)

Far right: Map used by Captain Jacober on mission against Hanoi. Stars indicate SAM sites. (Courtesy Captain Jacober).

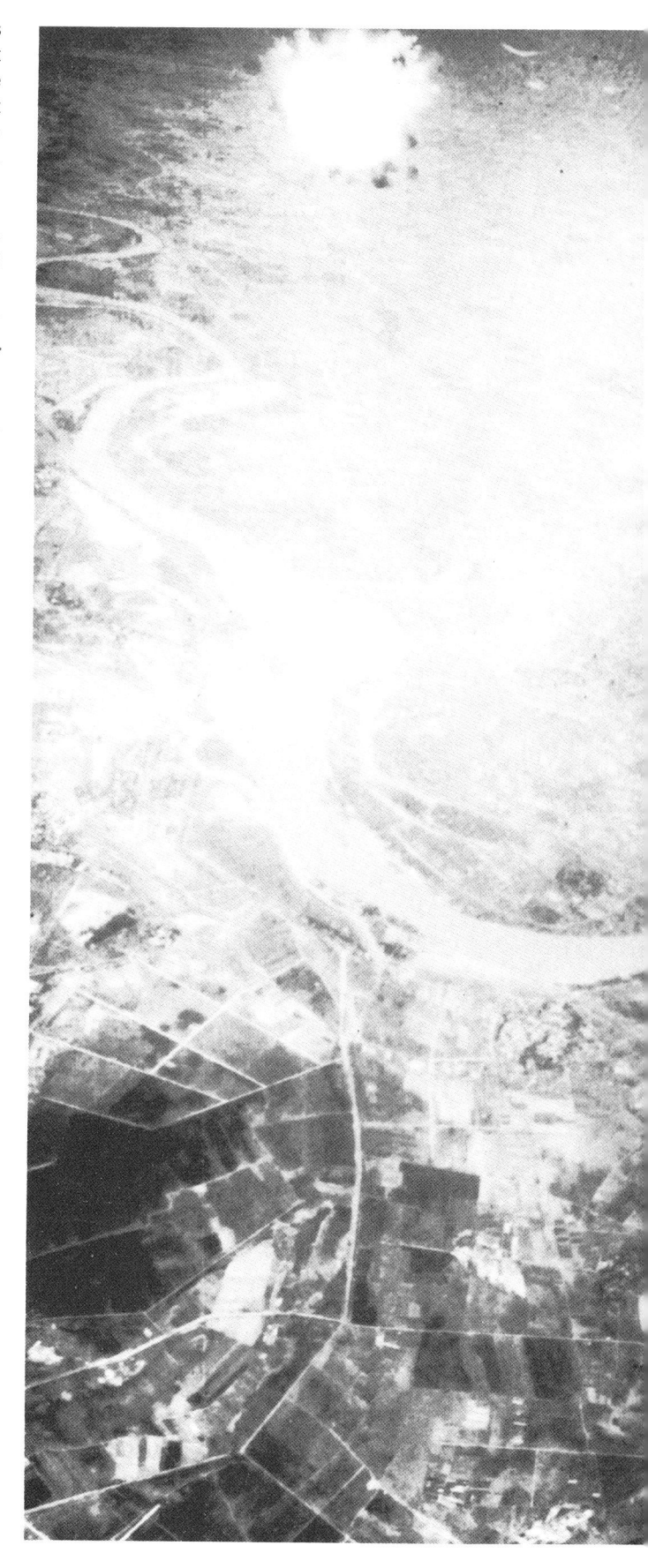

(VV) R2
HAI PHÒNG
NAM ĐỊNH
THUAN NGHIEP
NINH BÌNH
THANH HÓA
VINH
ĐỒNG HỚI
ASIAN COASTAL BUFFER ZONE
(VV) R2
INCOMPLETE
(VTD46)
(VTR42)
20°
19°
18°
17°
105°
106°
107°

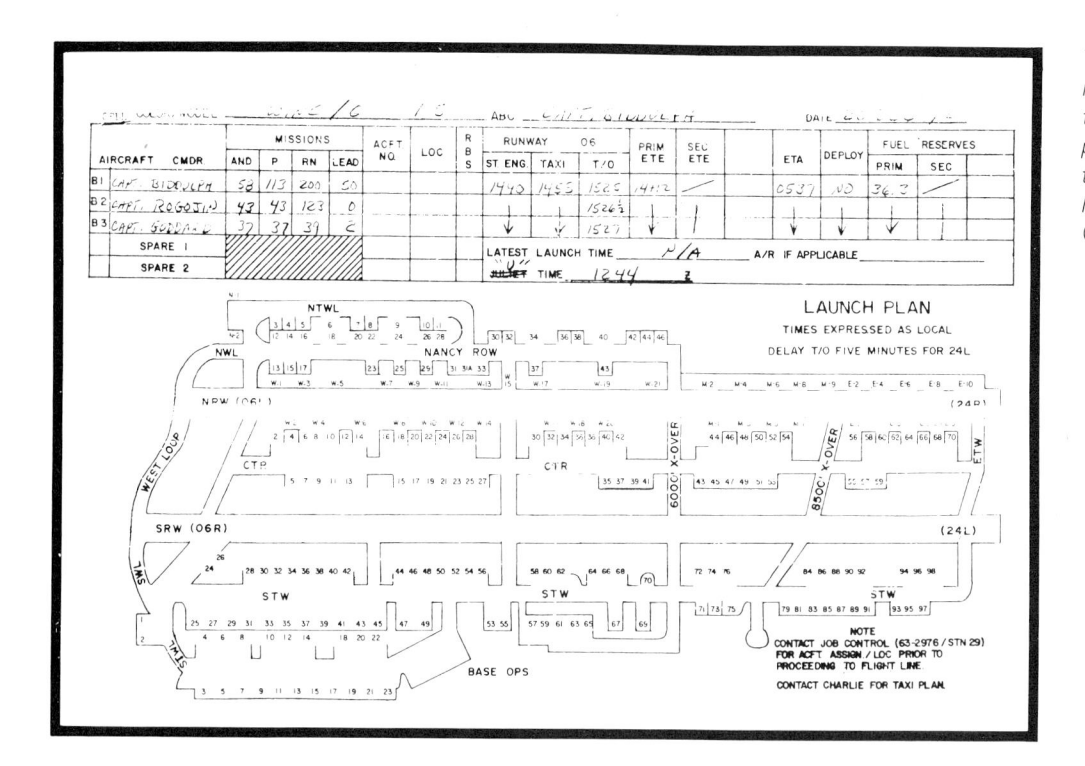

The sheer number of aircraft made it necessary for all crews to be issued a specific launch plan showing engine start up times, launch times, parking locations, etc. (Courtesy Captain Jacober)

defended territory in history, dropped 15,000 tons of bombs, and sustained 15 losses, or less than 2% of the sorties. About 1,240 SAMs had been fired against them.

By the eleventh day, the North Vietnamese were at the mercy of the B-52; had the bombing continued there would have been little or no SAM opposition. Then, instead of continuing to pound North Vietnam until it not only agreed to negotiate but agreed to surrender, the B-52s were returned to routine Arc Light missions. The last one came on August 15, 1973.

Eight years of Arc Light had seen 126,615 B-52 sorties; the B-52 had grown from a desperation weapon, thrown in when there was nothing else, to become the final instrument of the war.

Within a few months the North Vietnamese would ignore the conventions signed in Paris, and come again down the peninsula until they owned it all. Saigon became Ho Chi Minh City, and operations were resumed in Cambodia, Laos and Thailand. The only thing that can be certain is that they have not forgotten what the B-52 and its "iron bombs" can do.

Mention must be made of the attitude and morale of the crews which fought this hard battle, many of them finding it the toughest and most harrowing period of the long years of the war. Unlike the B-17 crews which came to England in World War II fully supported by public sentiment, younger, less knowledgeable, and certainly far less able to assess the danger, the B-52 crews were typically more mature, fully aware of the controversy surrounding the war and their part in it.

The years of combat had induced great strains on the family lives of the crews; the continual absences, the inevitability of the return to Southeast Asia, the sense that the war was endless caused frustration and bitterness that was reflected in the crew behaviour.

And yet despite all of the adverse conditions, it was the crew unit itself which kept individuals going. There are hundreds of cases of crew members not reporting sick so they wouldn't be taken off flight status and miss a mission with their crew. Many of the individuals had opportunities for other jobs as a result of promotions and so on, but would elect to stay with the crew. And in the heat of combat, it was this sense of loyalty and belonging which prevailed and which led to the superbly professional results.

There were many unusual side effects, too. As the crew members became veterans of hundreds of missions, they acquired a self assurance that enabled them to bend the iron rules of SAC. The pilots would set up special frequencies, not monitored by Bomber Control, and would chat on the way back. One group even played Bingo—the church social kind—on the long return journey. On the ground, discipline relaxed somewhat, with the crew members secure in the belief that the worst thing that could happen to them would be to be sent home—and they knew they were too valuable for that. Other changes in dress could be noted; one affectation was specially tailored, sometimes outrageously decorated "dress flying suits", some looking like they were designed for a British colonial planter, some for Reichsmarschall Göring, but all in a sort of gentle put-down of the strange way of life they were forced to lead.

The change has been maintained to some degree in SAC—the leavening of mission-hardened veterans has brought about a certain flexibility in outlook, a certain respect for individuality that did not exist before. The phenomenon of maturity taking liberties is not uncommon—RAF pilots leaving their top blouse button undone, the 50 mission crush hat—but it was unusual, and quite healthy, for SAC.

NEW PLANES FROM OLD

No aircraft company has ever provided the assiduous follow-up service that Boeing has provided the USAF on the B-52, and no aircraft company has ever sold the same airframe two or three times to the government, each time modified for a new role, or strengthened for additional service life.

The last statement sounds almost facetious, but it is a fact; Boeing did so well in the original design of the B-52 that in later years, when replacement aircraft would ordinarily have been required, none could do the new mission at less cost than a modified B-52.

The reason that Boeing has been able to provide the programmed modifications stems in part from the company's organizational set up and in part from the spirit of service which permeates it.

The B-52 originated in Seattle, in the minds of some of the best bomber men in the business. As noted, all B-52s from the XB through the C models were manufactured in Seattle. In late 1953, however, the Air Force decided that it wanted a second source for the aircraft. US thinking always preferred inland plants for aircraft manufacturers, and nowhere is more inland in the US than Wichita, Kansas. Furthermore, B-47 production was scheduled to end, and the fine labour force built up in Kansas needed to be kept occupied.

About 150 to 200 key Boeing personnel were transferred from Seattle to Wichita to aid in the transition of manufacture from B-47s to B-52s. One of the principal players was Beverly W. "Bev" Hodges, who became a mainstay of the programme, with an encyclopedic knowledge of the aircraft and the people behind it. "Bev" has helped immensely with this book, using his photographic memory to recall principals, events and documents.

The B-47 phase-out and the B-52 phase-in were nicely timed so that engineering and manufacturing personnel were able to move from one programme to the other, and the first B-52D to fly was built in Wichita. Seattle and Wichita shared B-52D, E and F production, but all G and H models were built in Wichita only.

It was obvious early on that a finite number of B-52s were to be built, and while some routine modifications were expected, no-one anticipated that the venerable bomber would become a career, a way of life, for many of the employees.

The Boeing-Wichita group was not passive about prospective B-52 changes. Early in 1959 an Operations Analysis unit was set up to analyze B-52 effectiveness against Soviet threats, and to determine exactly what could be done with the B-52 to improve it. In the course of the last 20 years,

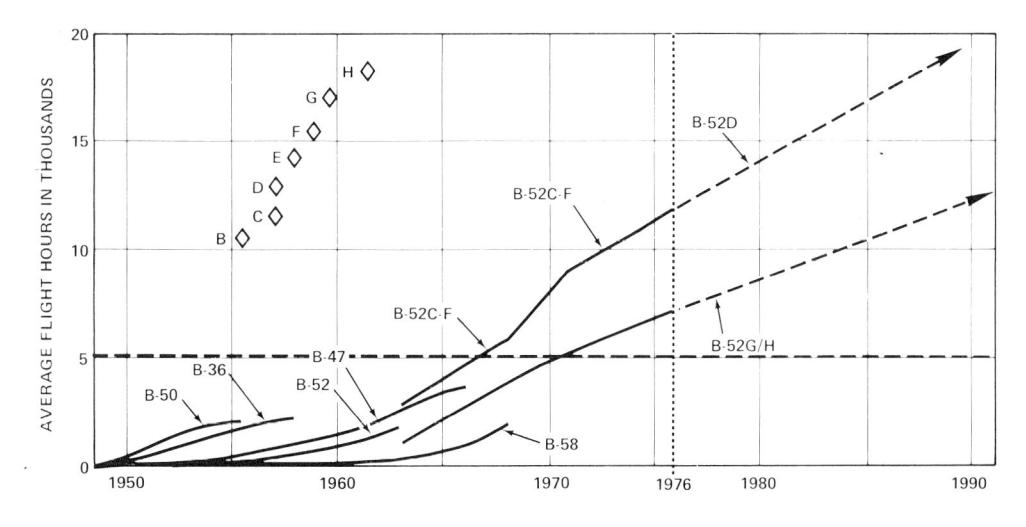

◇ DELIVERY DATE OF FIRST B-52 MODEL INDICATED

The utilization of bombers in the past was always for a relatively brief period of time, because they became obsolete before they wore out. As a result fatigue and corrosion control, while important, were not of the overwhelming importance that they became to the B-52, which almost inadvertently came to have an indefinite service life. The chart points out that while the B-50, B-36, B-47 and B-58 had relatively brief careers, with relatively few flying hours, the B-52 has served much longer and flown many more hours. It is projected for use for what would have been an unimaginably long period for any other bomber. (Boeing)

Boeing-Wichita has evolved system improvement concepts and capabilities which were demonstrated to Air Force leaders as means to keep the B-52 viable as a weapons system. The time period for work like this is critical; you must determine what is needed sufficiently in advance of the threat to build a demonstration model, sell it to the Air Force, have the Air Force obtain the funds and approval for its adoption, and then produce and install it in time for it to be effective.

The Operations Analysis unit has done this time after time. Now headed by Helene K. Little, a brilliant bright eyed woman with a computer like mind, the organization is currently working at least a decade in the future to see how long the B-52 will have a useful life.

Mrs. Little has contributed much to the development of methods and techniques used in effectiveness evaluation of electronic countermeasures from the time they were little more than noise generators to the ultra sophisticated equipment being installed today. The OA shop is run on the lines of a War Room, and various nuclear scenarios are played out in terms of known or possible conditions. The USAF does similar war gaming, and likes to compare the Boeing-Wichita studies for verification.

Early in the unit's life, interest was centred on threats and targets; in recent years, emphasis has also been put on evaluating the impact of on-board equipment reliability on mission effectiveness.

The same sort of concern and innovation is pervasive in the other Boeing organizations, and has permitted the company to undertake structural modifications of a size and scope that would not have been countenanced in aircraft in other days. The basic William Allen (former Chairman of the Board of The Boeing Company) philosophy that "What is good for the Air Force is good for Boeing" prevails in all units, and has even resulted in Boeing recommendations against desired Air Force projects that would have made money for Boeing, but which Boeing did not think would be in the interest of the service.

There is a healthy interaction between Boeing and the various Air Force organizations it serves. The company works with the Aeronautical Systems Division of Air Force Systems Command, at Wright Patterson Air Force Base, Ohio, during the development of new equipment. After the major production stages, the focus shifts to the Air Force Logistic Command depots. These are now designated as the Oklahoma City Air Logistics Center (OCALC), the San Antonio Air Logistics Center (SAALC) and the Warner Robins Air Logistics Center (WRALC).

The term so frequently bandied about, "military industrial complex", takes on a new and finer interpretation when applied to the Boeing/Air Force B-52 programmes. It is a cooperative relationship through the period of selection and procurement of a particular modification, and a competitive relationship when it comes to negotiating costs. Boeing makes a high-quality product and charges for it. The Air Force demands high quality, but will not pay more than its audit teams determine to be fair. It works out to be a satisfactory arrangement for both parties.

The Problems and The Challenges

In the course of its long life, the B-52 has been the centre of many problems, some generated by such routine facts of life as corrosion and fatigue, some by design errors, but mostly by changes either in the environment in which it was to operate, or by the need for a greater capability. For convenience, we will approach the B-52 modification programme on two parallel lines.

The first will deal with structural modifications for fatigue, caused either by normal wear and tear or by the new conditions under which the aircraft had to operate.

The second will deal with equipment modifications which were required either to meet the challenges of these new conditions, or to extend the aircraft's capabilities.

Structural Problems

As the accompanying chart shows, the B-52 has experienced, and still projects, a structural life far beyond any previous bomber. The very nature of the B-52, and its original projected life of about 5,000 flying hours, was due to the design criteria set up to achieve the desired performance. The aircraft had been designed from the start to have a very low empty weight to design gross weight ratio. This meant that the engineers had to design in just enough strength to do the job so that it would achieve the extraordinary range and speed desired. This was done in part by reducing the permissible limit load factors to 2·0 gs, a figure that was reduced even further for certain operational conditions.

Yet fate and time intervened, and instead of having about the same service life as the B-29 or the B-47, the B-52 suddenly found itself being used year after year, with no projected retirement date.

As a result, there were modification programmes early on to extend the structural service life of the aircraft. A typical one took place in 1957-59, and was called the "Sunflower" programme, a tribute to Kansas, the Sunflower State. B and RB-52B aircraft were modified to have the same combat capability of the first Wichita built B-52D aircraft. A complete IRAN (Inspect and Repair as Necessary) was performed, and more than 150 kits installed to improve offensive and defensive capability, crew comfort, the fuel system, hydraulic system, wing trailing edge, communications and other miscellaneous areas.

Besides formal programmes like "Sunflower" there were a continuous flow of changes that occur in service and which were a part of the normal Air Force updating process.

These changes can range from altering the wording on the decal next to the electric hot cup to a complete structural re-work due to a fully fledged approved Engineering Change Proposal (ECP). More than 1,800 Engineering Change Proposals have gone through the B-52 Configuration Control system, and have varied from minor proposals for a weight reduction to full scale ECPs which we will follow in assessing the "stretch" of the aircraft.

It would probably be well to discuss in layman's terms what an ECP is, for we will be covering hundreds of millions of dollars worth. It is necessary for configuration control to be maintained over weapon systems. One must know what modifications have been made to the system to know its capability, to be able to supply spare parts intelligently, and to be able to repair it. There have been occasions in the past when insufficient configuration control was exercised over a weapon system. (The Republic F-84 aircraft and the Atlas missile come to mind.) The operators in the field and in the maintenance depots were presented with engineering nightmares. They literally did not know what the system consisted of, and there were enormous problems in terms of possible malfunctions, the emission of frequencies which might have a catastrophic effect on other equipment and the simple inability to make things work.

Therefore a "baseline configuration" is determined and documented for the aircraft and any changes to this baseline must be made a matter of record, usually by an Engineering Change Proposal.

ECPs may be of varying classes, depending upon their importance, and can originate either with the Air Force or the contractor. Once originated, they must be approved by the appropriate Air Force Configuration Control board, and entered into the system. The effects on weight, drag, electronic emissions, space and so on are carefully noted, and all drawing changes are carefully controlled. The ECPs must also be budgeted for, and cost effectiveness evaluations are made. Because money for modifications is so hotly contested for, only a few of the many ECPs proposed are approved.

Once an ECP has been approved, it can be incorporated in a variety of ways. The easiest is when the ECP approval occurs prior to the production of the aircraft, so that the change can be incorporated in the production line. If it occurs after production, it may be done in the field, by means of a kit supplied by AFLC, if it is relatively simple. If it is relatively complicated, but not of immediate urgency, it may be accomplished during the next routine maintenance cycle. If it is extremely important it may be done at special depots set up for the task, or by a massive fleet recall programme to the various Air Logistics Centres.

Structural problems are heralded in a variety of ways. The most urgent is when an accident occurs. SAC and AFLC had been agonized by the nature and frequency of fatigue problems, particularly the notorious B-47 "milk bottle fitting", and began with the B-52 a number of programmes intended both to anticipate and forestall the issue.

The first and most obvious of these was increased inspection and maintenance of probable trouble spots. Unfortunately, in an aircraft the size of the B-52, there are literally thousands of potential trouble spots. Any place where two pieces of metal are joined by a fastener, any place where there is a cyclic alteration of stress, any place subject to vibration can result in fatigue, corrosion and ultimately failure.

In addition to increased inspection, SAC and AFLC began a "lead the fleet" programme in which certain selected aircraft were flown much more intensively than the fleet average, often more than 125 hours per month, so that 1,000 to 1,500 hours flying time could be accumulated in excess of the normal usage. Logically fatigue would show up in the high-time aircraft first, but it did not always work that way.

These measures were reinforced by a series of cyclic tests, in which aircraft would be placed in special jigs and subjected to a series of artificially induced stresses over a long period of time to see if failure could be induced.

The AFLC also initiated a fleet damage monitoring programme in which aircraft were fitted with equipment that allowed tracking by individual aircraft of the status of the structural integrity of the fleet.

By 1964 there was sufficient evidence of fatigue problems to launch the B-52 Hi-Stress Modification programme. This was a $62·9 million effort involving 32 separate ECPs which modified individual areas of the aircraft. For example, in the B-52A through F models, eight ECPs covered small areas on the upper surfaces of the wings, while 16 of them covered similar small areas on lower surfaces. Doublers would be applied as reinforcements, new larger extruded angles would be substituted for original structural members for increased strength, cracks would be repaired, internal reinforcing plates were applied to surface splices, and so on. It was a difficult, time consuming task, not requiring a lot of extensive jigging, but needing a high degree of skill on the part of the workers to ensure quality workmanship in the installation and scrupulous care in the inspection of the work. The problem in repairs like this is that, improperly done, they can do more harm than good.

As time passed, the environment in which the B-52 operated changed from high to low altitude, and further modifications became necessary. It was discovered, as a result of three tragic crashes*, that the main bulkhead in the aft end of

Below: *The safety record of the B-52 compares very well with other bombers, particularly in its later years of service.* (Boeing)

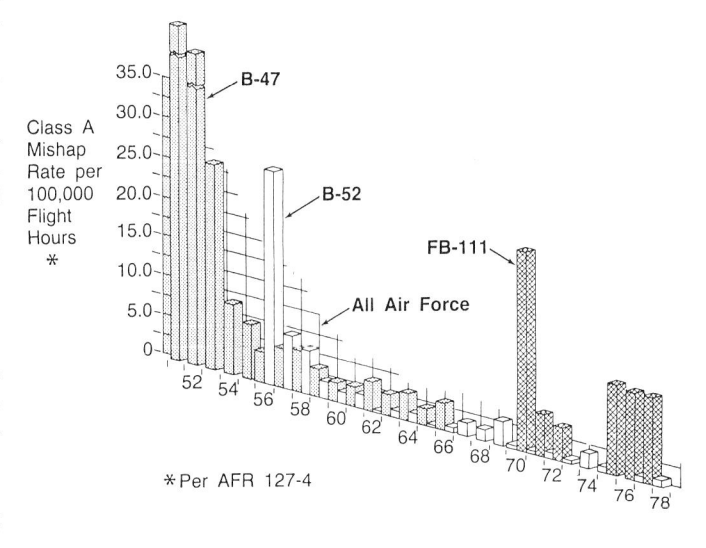

*Per AFR 127-4

---

*B-52B, 53-390, lost at Monticello, Utah, January 1961; B-52C, 53-406, lost at Greenville Maine, January 1963; B-52E, 57-018, lost at Mora, New Mexico, January 1963. All were lost due to severe atmospheric turbulence, which exceeded the gust velocities for which the aircraft were designed. 53-390 and 57-018 were at high altitude, and 53-406 was at low altitude.

ECP 1128
CCP 1185
ECP 1124

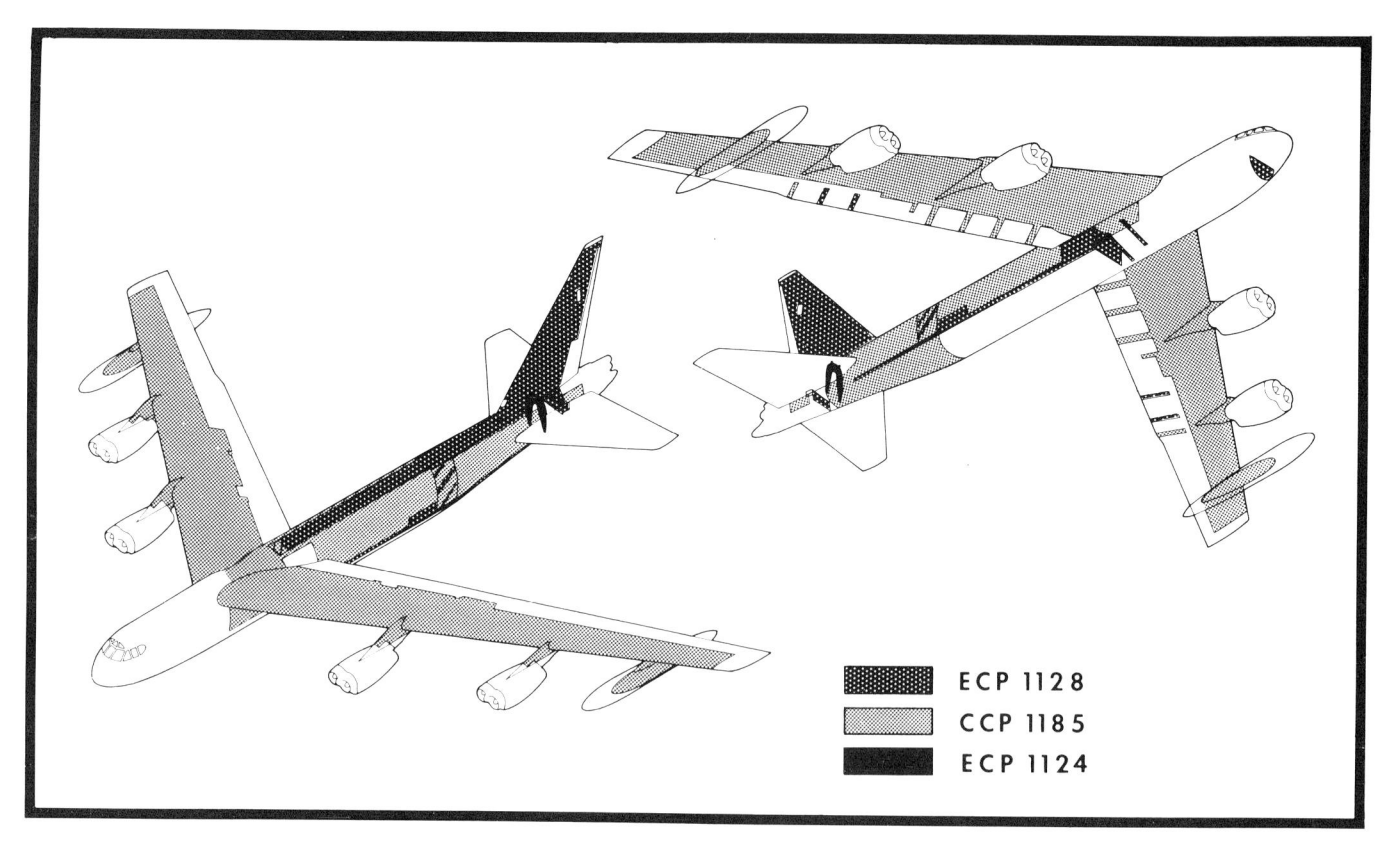

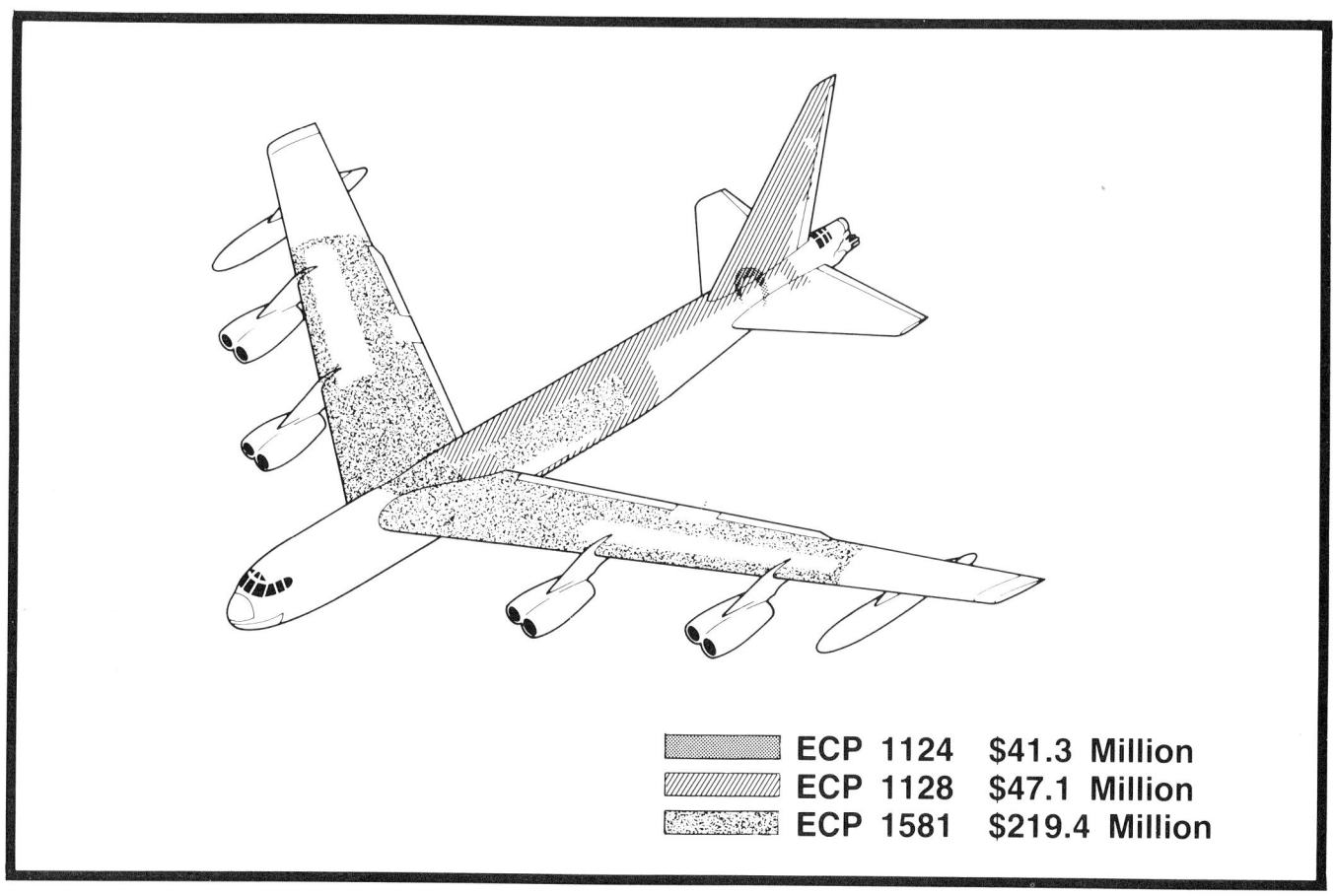

ECP 1124 $41.3 Million
ECP 1128 $47.1 Million
ECP 1581 $219.4 Million

the fuselage had insufficient strength to sustain the loads induced by severe turbulence.

ECP 1124 was created at a cost of $41·3 million to fix the bulkhead problem. Shortly thereafter, ECP 1128 at $47·1 million strengthened the upper fuselage and vertical fin.

"Pacer Plank"

By careful monitoring of both fleet and individual aircraft condition, the Air Force was aware in 1972 that it faced a critical fatigue problem with regard to B-52Ds that had served so valiantly in Vietnam as "iron bomb" carriers.

A series of conferences with the best structural experts in the service resulted in a host of recommendations that would come to be known as Project "Pacer Plank", ECP 1581. The first step in the programme was to restrict the operating conditions of the B-52D fleet so that the possibility of undetected cracks would not cause catastrophic failures. Flight hours, speed, load carried and limiting load factors were all drastically reduced, and a rigorous programme was initiated to choose the best 80 aircraft from the remaining fleet of 128 B-52Ds. SAC's maintenance records were investigated to determine for each aircraft the total number of flight hours, number of low level and refuelling hours, total number of sorties and if any battle damage had been incurred.

A partial B-52D airframe was subjected to destructive tests at the Wichita facility to determine what was the actual residual strength of wings with cracked panels. Saw cuts were made at various locations in the lower wing skin to simulate cracks. The airframe was then inverted and placed in a special fixture. The wing was instrumented and then loaded with shot bags until it collapsed.

To further assure continued safe operations until the structure could be modified, the B-52Ds were then subjected to a proof load test. Each was placed in a special fixture, and 100% design limit load was applied to the wing in both up and down directions. No failures occurred, and the proof tested aircraft were returned to service until they could be modified.

The structural modifications then involved the virtual rebuilding of major portions of the wings and large portions of the fuselage. The wing modifications included redesigned lower skins, spar webs, stiffeners, rib chords, inboard upper skins and stiffeners as well as new leading and trailing edge

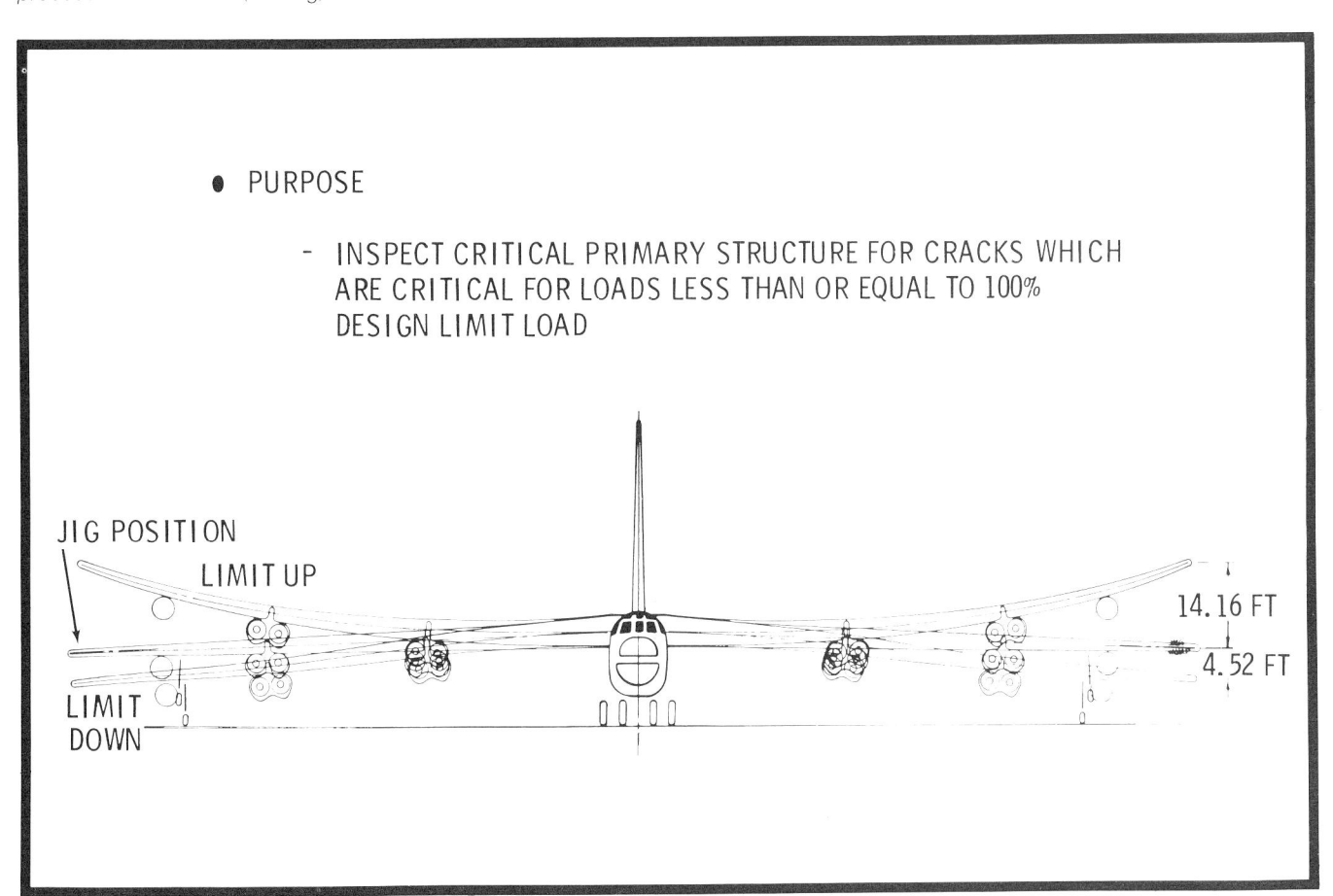

Left: *The sheer amount of metal removed and replaced from the B-52 fleet over the years is incredible. This plate shows the areas replaced or strengthened by ECP 1185. (CCP in this instance means Contractor's Change Proposal, which when approved becomes an Engineering Change Proposal).*

Below left: *The B-52D had additional work done to it, as shown.* (Boeing)

Below: *Actual testing forced aircraft through a stiff selection process. None failed.* (Boeing)

- PURPOSE
  - INSPECT CRITICAL PRIMARY STRUCTURE FOR CRACKS WHICH ARE CRITICAL FOR LOADS LESS THAN OR EQUAL TO 100% DESIGN LIMIT LOAD

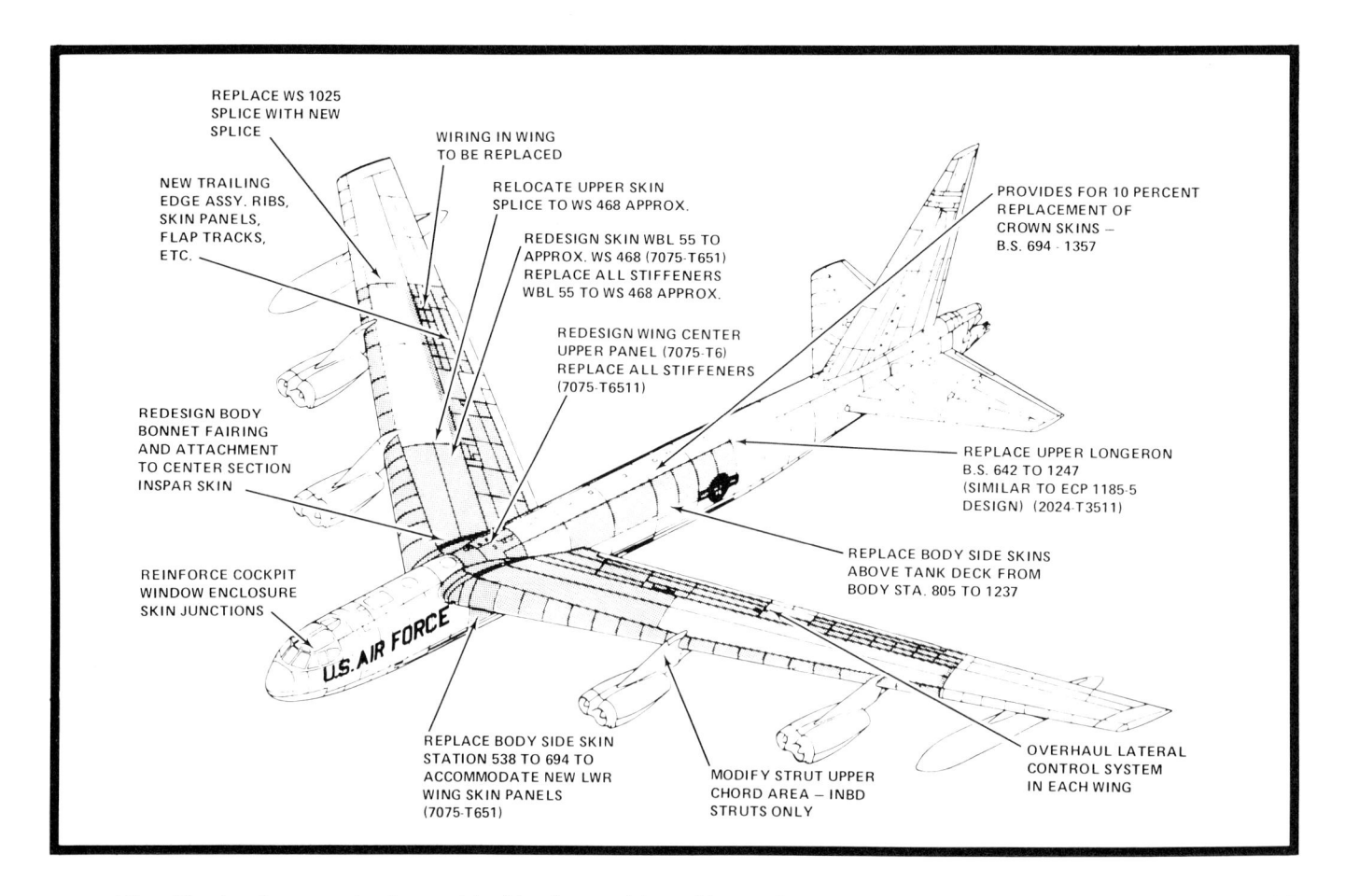

assemblies. The fuselage received new side skins from midway in the bomb bay to the aft bulkhead of the aft wheel well.

Each aircraft was partially disassembled, with more than 8,000 separate components being removed before proceeding to the modification line. Complete inspection was given for corrosion and depot level maintenance. While the aircraft was disassembled, numerous other non-structural changes were incorporated, to improve such things as ejection seats, and the maintainability of several sub-systems.

When "Pacer Plank" was finished the Air Force had a B-52D fleet with a service life which could extend to the year 2,000, for a total cost of $219·1 million.

The problems which appeared on the later model B-52Gs and Hs were similar to those encountered by the D models, with one major exception.

In the desire to reduce the weight of the structure when the Gs and Hs were being designed, an advanced aluminium alloy, numbered 7178, was selected for use in the lower wing because of its higher strength to weight ratio. Unfortunately, the durability and damage tolerance characteristics of this alloy were not completely known, and it soon became apparent that the B-52G and H wings would not have an adequate structural life. The potential scope of the problem was highlighted by the crash of a B-52G, 58-187, at Seymour Johnson AFB, South Carolina. The aircraft, which had only about 650 hours flying time, had developed a massive fuel leak at altitude with JP-4 pouring out of the lower wing skin. The pilot brought the aircraft back in for a landing, but when

Above, above right and right: ECP 1581, called "Pacer Plank" was very extensive. (Boeing) Reinstallation of aged tanks was a terrific task. (Boeing)

full flaps were extended, the change in stress caused the wing with the broken panel to fail, and the aircraft went in.

It was necessary to completely redesign the wing, reverting to the more durable and damage tolerant 2024 aluminium alloy in ECP 1050, at a cost of $139·1 million.

There were additional structural modifications. ECP 1185-5 extended the service life of the fuselage at a cost of $65·2 million. ECP 1128-1 strengthened the fuselage and vertical fin, for $87·9 million, and the service life of the fuselage was extended by ECP 1185-5 at a cost of $65·2 million.

All of these modifications were to extend the service life of the aircraft. There were also actions taken to reduce stress. All aircraft have inherent characteristics which determine the quality of their flight during operation in turbulence and when manoeuvring for terrain avoidance at low altitudes, in terms of vibration, responsiveness and so on. The Convair B-58, with its delta wing, had good low level ride characteristics. The B-52's ride characteristics, to put it mildly, were awful. The aircraft responded to turbulence with a severe jolting ride that would sometimes shake the pilots so much that the instruments could not be read. In early models, the poor tail gunner, 150 feet behind the pilots, had an even worse time, and the severe buffeting sometimes resulted in

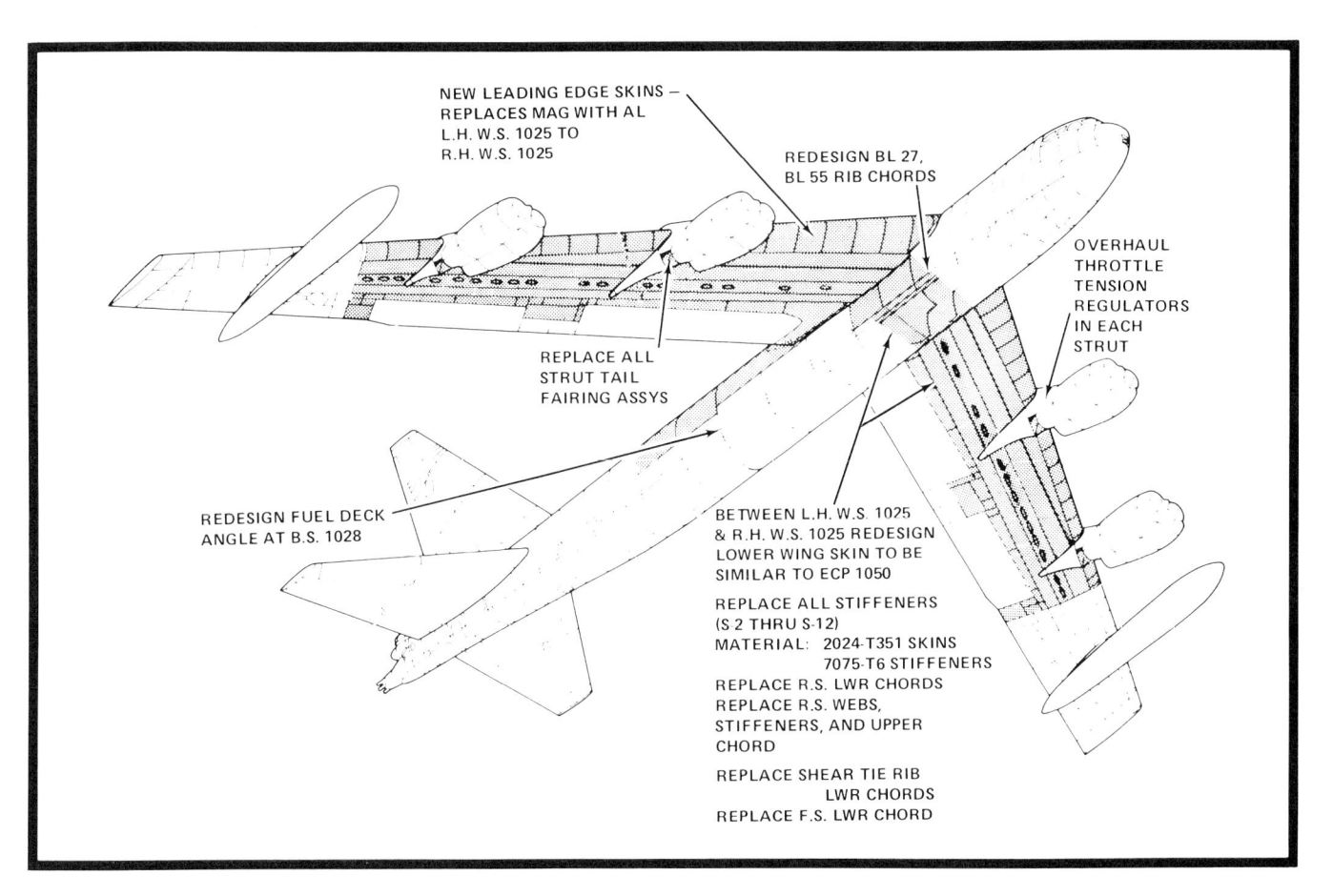

Top figure labels:

- NEW LEADING EDGE SKINS — REPLACES MAG WITH AL L.H. W.S. 1025 TO R.H. W.S. 1025
- REDESIGN BL 27, BL 55 RIB CHORDS
- OVERHAUL THROTTLE TENSION REGULATORS IN EACH STRUT
- REPLACE ALL STRUT TAIL FAIRING ASSYS
- REDESIGN FUEL DECK ANGLE AT B.S. 1028
- BETWEEN L.H. W.S. 1025 & R.H. W.S. 1025 REDESIGN LOWER WING SKIN TO BE SIMILAR TO ECP 1050
- REPLACE ALL STIFFENERS (S-2 THRU S-12)
  MATERIAL: 2024-T351 SKINS
  7075-T6 STIFFENERS
- REPLACE R.S. LWR CHORDS
- REPLACE R.S. WEBS, STIFFENERS, AND UPPER CHORD
- REPLACE SHEAR TIE RIB LWR CHORDS
- REPLACE F.S. LWR CHORD

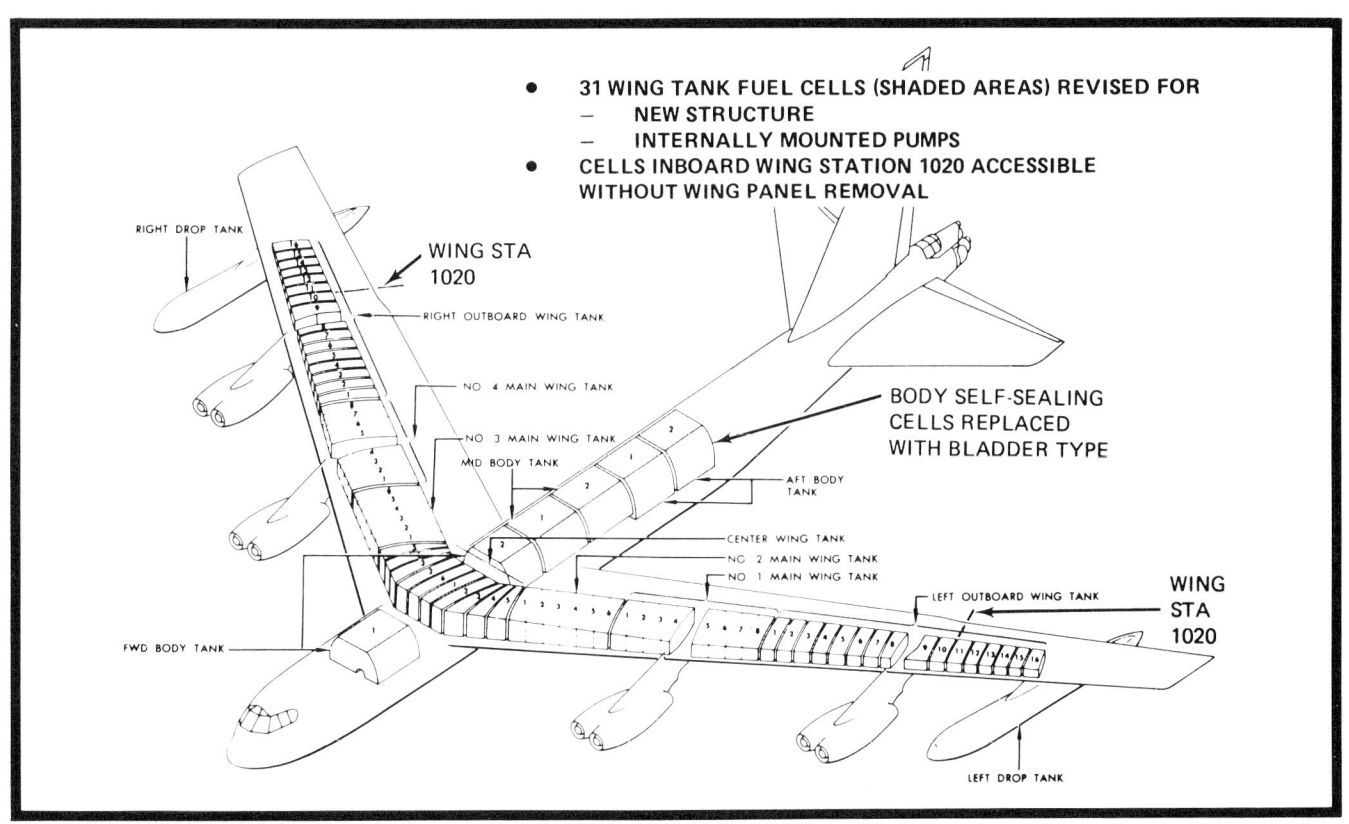

Bottom figure labels:

- 31 WING TANK FUEL CELLS (SHADED AREAS) REVISED FOR
  - NEW STRUCTURE
  - INTERNALLY MOUNTED PUMPS
- CELLS INBOARD WING STATION 1020 ACCESSIBLE WITHOUT WING PANEL REMOVAL

- RIGHT DROP TANK
- WING STA 1020
- RIGHT OUTBOARD WING TANK
- NO. 4 MAIN WING TANK
- NO. 3 MAIN WING TANK
- MID BODY TANK
- BODY SELF-SEALING CELLS REPLACED WITH BLADDER TYPE
- AFT BODY TANK
- CENTER WING TANK
- NO. 2 MAIN WING TANK
- NO. 1 MAIN WING TANK
- LEFT OUTBOARD WING TANK
- WING STA 1020
- FWD BODY TANK
- LEFT DROP TANK

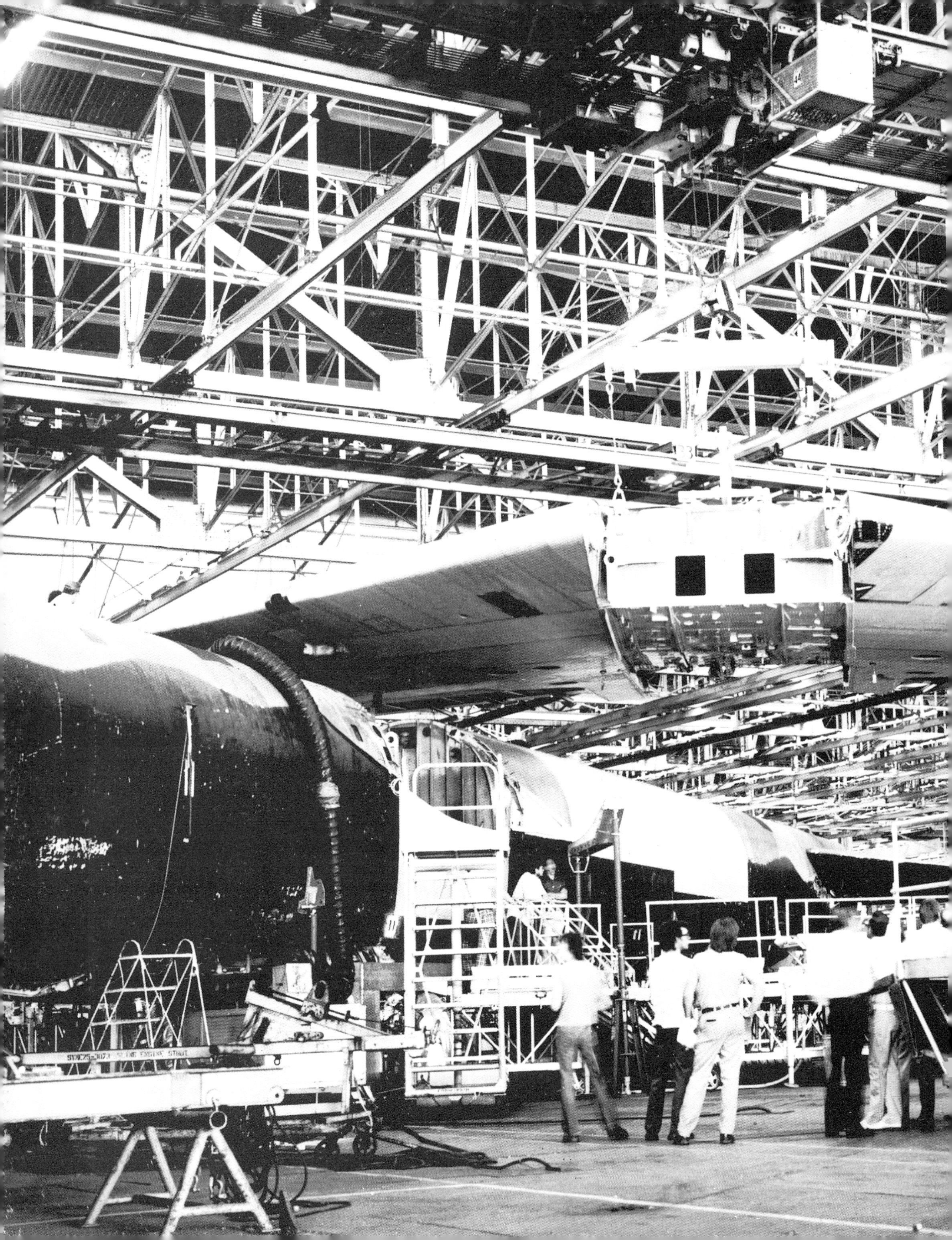

Above: *The wear and tear is indicated by the chipped paint, dings and so on. While in process, the aircraft received an extensive equipment updating.* (Boeing)

Left: *A critical moment, when the extensively rebuilt and reskinned wing rejoins the structure. Contrast the almost new appearance of the wing with the weather worn fuselage.* (Boeing)

injuries and often in airsickness. Similarly, the radar navigators and navigators, cowering in their dark cubby hole, would be thrown about so much that it was dangerous for them to loosen their seat and shoulder harnesses.

The tail gunner had it easier in the B-52G and H models since he was located up forward with the rest of the crew. The G and H models, ironically, were even more vulnerable to the effects of turbulence because of their shorter fins, and because they had dispensed with ailerons in favour of spoilers. They were more prone to Dutch Roll, and a pilot getting out of synchronization with the motion could inadvertently amplify the Dutch Roll and overstress the aircraft.

Boeing and Air Force engineers devised a Stability Augmentation System (SAS) to improve the low level flight characteristics, reduce structural loads, and improve controllability in turbulence. The system utilized inputs from the new pitch and yaw sensors and an accelerator unit, which were integrated in a computer and delivered to the Stability Augmentation System electronic control units. Hydraulic actuators were added for the rudder and elevator, and the resulting "power steering" greatly alleviated both crew work load and structural fatigue.

The need to revise structural design gust load criteria to recognize the existence of infrequent but very severe turbulence that can occur during operation over mountainous

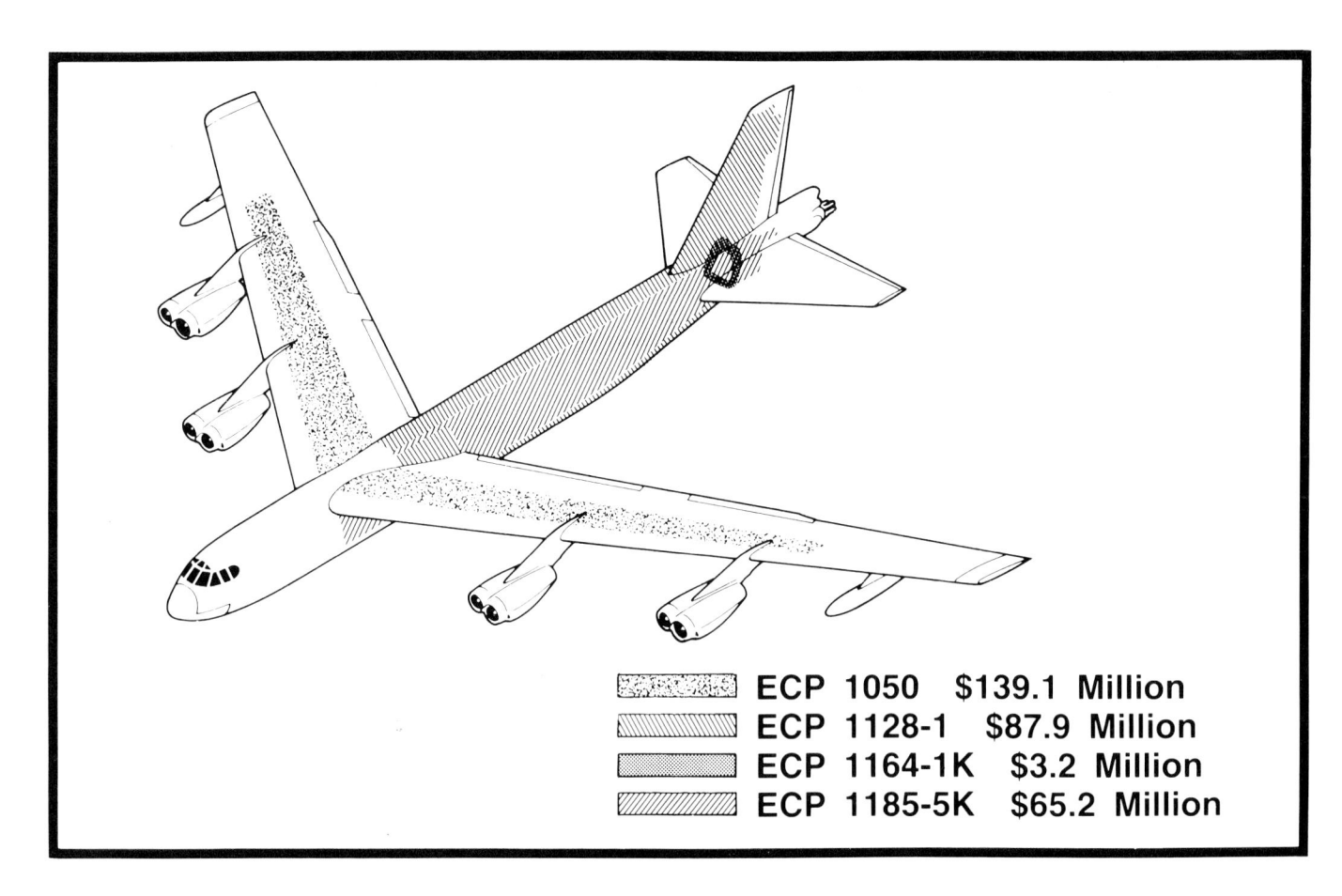

Pattern	ECP	Cost
(dotted)	ECP 1050	$139.1 Million
(diagonal lines)	ECP 1128-1	$87.9 Million
(gray)	ECP 1164-1K	$3.2 Million
(hatched)	ECP 1185-5K	$65.2 Million

terrain in the presence of high surface wind velocities, was made manifest in the series of accidents which had literally torn B-52s apart. The first occurred in 1959 at Burns, Oregon, where a Boeing crew in B-52D 56-591 crashed.

Boeing-Wichita instrumented a B-52 and embarked on an extensive investigation of the problem. The Boeing crew, headed by test pilot Charles F. "Chuck" Fisher, was engaged in an eight hour flight which would test indicated airspeeds of 280, 350 and 400 knots over mountainous terrain at 500 feet altitude above the ground.

Above: *The B-52G and H had rework on an extensive scale. (Boeing)*

Below: *The crew was reluctant to abandon the aircraft because the instrumentation installed was sure to have valuable information on the effects of clear air turbulence, so speed was reduced to 210 knots indicated air speed and the airplane was flown, gingerly, to Blytheville AFB, Arkansas, where the wind was straight down the runway, and an approach could be made over an unpopulated area. Fisher got the plane on the ground safely. (Boeing)*

During the test the turbulence became so extreme that the crew decided for safety's sake to discontinue the mission and climb to 14,000 feet. There a sudden five second blast of clear air turbulence moved the aircraft sharply to the right, sideways through the air, and broke off the vertical fin. Fisher stated that it felt like a severe sharp edged blow, followed by several more. The aircraft developed an instantaneous roll to the left at a high rate, the nose swinging up and to the right, rapidly. Fisher reduced power to idle, and the aircraft then rotated nose down.

Fisher applied full opposite controls, without apparent effect; the rudder pedals were locked. Applying airbrakes, he got the airspeed down to 210 knots. With 80 degrees of wheel deflection, he managed to get the aircraft stabilized in somewhat level flight.

The next three hours involved some very harrowing, very sensitive flying. The crew knew how important it was to get the specially instrumented aircraft back safely, so that the data of an incident which had caused other crashes would be preserved. On the other hand, they had no idea if the damage would propagate, throwing the aircraft out of control, or what the effects of lowering the gear or flaps would be, or even what effect the consumption of fuel might have.

They had made an immediate control to Denver Flight Service, notifying them of the emergency. The co-pilot, Dick Curry, had lowered his seat in preparation for ejection, if necessary, and he could not see the Hound Dog missile when he glanced out of the side window. Assuming it was missing, he notified Denver Flight Service, which became very agitated at the thought of a nuclear missile being lost. (It was inert, of course.)

Fisher had contacted Wichita immediately, and the usual staff of experts was assembled. Long time test pilot Dale Felix was scrambled in a North American F-100 to act as chase, while a KC-135A tanker prepared to get airborne.

Felix joined up with the B-52 and surveyed the damage. A quiet, reflective man, he had not yet made any comment when he heard Fisher remark "we've slowed down to 220 knots, we're stable, and I'm going to handle it pretty carefully".

Felix cut in with "Chuck, that's a good idea." Fisher responded "What do you see?" Felix said "All of your rudder and most of your vertical fin are gone."

There was a silence, and Fisher said "Don't I even have 50%?" Felix replied "No, you don't have 50%". There was actually only about 15% of the fin remaining.

After a lot of consultation, it was decided to lower the aft main gear, to provide some lateral directional stability. A number of other steps were taken, including transferring fuel to move the centre of gravity forward, and to deploy only the outboard airbrakes.

The entire crew remained ready to eject, but Fisher and Curry decided that they could land the aircraft under favourable conditions. A quick check around the country revealed that Blytheville Air Force Base, Arkansas, offered an approach to landing over unpopulated areas, and the wind was straight down the runway. Fisher brought the aircraft in, flaps up, and in his words "the landing was not my best one but the airplane was drifting left off the runway and the only way to stop it was to get it on the ground."

The data from the instrumented aircraft showed where some improvements in structure might be made, but it showed clearly and in concert with other evidence that no aircraft could be made safe from the most extreme values of clear air turbulence. The best method remained to plan flights to avoid encountering it, and once encountered, to reduce airspeed, maintain attitude, and get out of the danger area downwind as soon as possible.

Meeting the Challenges

The Air Force's decision to retain a fleet composed of only B-52Ds, Gs and Hs simplifies a review of the extensive major modifications that have been installed to meet the challenge of operating in new environments. For simplicity's sake only the modifications made to the D series are considered first, then the one which affected all three models, and finally those applicable only to the Gs and Hs.

B-52D Modifications

Through 1965, all of the B-52s committed to "iron bomb" operations in Vietnam were B-52Fs. These aircraft could carry 51 750 pound bombs, 27 internally and 24 externally. A programme was set up to modify B-52D aircraft to carry a far greater load. Called "Big Belly" the modification extended the B-52D's internal capacity for 500 pound bombs from 27 to 84 or for 750 pound bombs from 27 to 42. It could carry 24 bombs, either 500 or 750 pounds, on the external pylon racks. The maximum bomb load was either 54,000 pounds, when carrying 108 of the 500 pounders, or 49,500 when carrying a full load of 750 pound bombs.

The modification programme not only increased the number of bombs that could be carried, but also provided for a "preload" system in which bombs were readied in the munitions area in special racks which could be clipped into the bomb bay in a minimum time. In actual practice, the bomb teams, sweating out their 179 days in the heat of Southeast Asia, became so proficient at the complex and dangerous task of building up and loading bombs that they could load the aircraft directly with almost the same speed as using the preloaded clips.

"Big Belly" cost $30·6 million and it added a thundering tactical artillery capability to SAC which had never been envisaged. It also added several armament options, including the use of mines and glide bombs. The D model also retained the capability to carry up to four nuclear gravity weapons as well.

Bomb/Nav System Improvements

The original bombing and navigation systems with which the B-52 began life were typical of the period, using vacuum tubes and requiring a great deal of maintenance. The years of hard usage, with the vibration and shocks encountered at low level and in refuelling were hard on the electronic equipment. Worse, from a supply standpoint, was the fact that much of it was long out of production, and spare parts could not be obtained.

The first bombing navigation system improvement, the addition of a low level capability by using terrain avoidance radar, has been discussed in Chapter Five. However, the total

Left: *The "Big Belly" modification to B-52Ds involved the maximum utilization of available space combined with the quickest methods of loading bombs. Here a pre-loaded clip is positioned. Boeing worked extensively with experienced USAF armament personnel, the "hands-on" people who knew how something should work in practice.* (Boeing)

array of equipment began to degrade, and an entirely new digital bomb navigation system was acquired at a cost of $126.3 million. The modification programme began in 1980 and it will delete obsolete analog computer equipment and replace it with modern digital units. A new inertial navigation system, new controls and displays and redesigned crew positions will enhance reliability and maintainability. As the equipment is installed, it will be "hardened" to withstand nuclear radiation.

Under the Department of Defense policy of maintaining the maximum amount of flexibility in its forces, the Strategic Air Command was given the responsibility to support the Navy, when called upon, to perform such roles as ocean surveillance, aerial mine laying and attacking surface vessels. The B-52D's "Big Belly" capability was improved by adding the GBU-15 glide bomb to its arsenal. The GBU-15 is a modular glide bomb with a television sensor that allows the B-52 to launch its weapon outside the lethal defence range of a ship. The optical sensor presents the radar navigator with a television picture of the scene forward of the weapon, so that he can guide it directly on target. The cost of retrofitting the B-52 with the necessary equipment, including a TV monitor, tracking control handle weapon system control panel, data link pod and data link control panel, was $5.2 million.

Shared Systems, B-52D, G and H

The $313.2 million dollar modification to retrofit the Advanced Capability Radar (ACR) terrain avoidance capability of the B-52 was described in Chapter Five. As effective as the ACR was, a new system was developed which would make it completely obsolete, the Electro-Optical Viewing System (EVS), which will be covered in the following section.

G and H Modifications

As the "youngest" and most capable aircraft in the fleet, and as the aircraft with the primary nuclear mission, the G and H models have received more sophisticated changes than has the D.

The 15 to 30 minute warning time of the launch of a Soviet intercontinental missile made it mandatory that SAC be able to get its fleet of nuclear bombers off the ground in as short a time as possible, from as many widely spread bases as possible. To assist in this, a retrofit programme to incorporate simultaneous start of all eight engines by means of cartridge starters was initiated in the mid-1970s. All G and H models were retrofitted at a cost of $35 million to improve pre-launch survivability by a few minutes.

EVS—Two Eyes in the Night

A major reason for the retention of the manned bomber in the US nuclear triad is the ability of the crew to make inflight decisions on the damage that has been done to previously struck targets and to determine whether another attack is warranted. Radar was intended for this use in the past, but it is not as accurate as needed, and it did not pick up the firestorms which would emanate from target areas. In addition, the low level penetration tactics increased the need for hazard avoidance capability. When operating in a nuclear environment, heavy curtains shutter the B-52 crews in even more tightly than bad weather. Most scenarios call for the B-52 penetrating a target area to be in "closed curtain" condition for almost the entire route.

Jack Funk, a long time Boeing engineer, and at the time of his invention the Chief of Flight Test, made the first experiments with what became known as the Electro-Optical Viewing System (EVS). Entirely on his own initiative he installed a Sony TV camera on the tail of a B-52 in 1964, during the course of an entirely unrelated test. It worked beautifully, and Funk soon communicated his enthusiasm to the Air Force representative Colonel Rick Hudlow. SAC was persuaded to issue a requirement in 1965 on the feasibility of using visual

Right: *The Advance Capability Radar (ACR) presented its terrain avoidance display on a 5 inch TV-like screen shown next to the attitude indicator on this instrument panel.* (Boeing)

Right: *The more modern Electro-Optical Viewing System (EVS) has a much larger display unit which not only shows the terrain ahead, but presents a "head-up" type display of other flight information via alpha numeric symbology.* (Boeing)

sensors, including infrared, in the B-52G/H fleet to improve damage assessment and strike capability and to complement the B-52 terrain avoidance system. A production contract was issued in 1970, after five years of experimentation.

The system, formally known as AN/ASQ-151 EVS, was created in a happy programme characterized by cooperation at all levels of the Air Force and the company. The technical problems were formidable, but the system worked well. It consisted of a Westinghouse steerable television sensor, capable of operating by starlight, a Hughes forward looking infrared sensor (FLIR), and interface equipment including a video distribution unit, a symbol signal generator, a servo control unit and monitors at four crew stations. A radar scan converter provides a connection between the EVS and the terrain avoidance radar, so that the radar video presentation can be displayed on EVS monitors.

The two pilots have EVS displays which show the terrain avoidance profile trace, a means to select between the

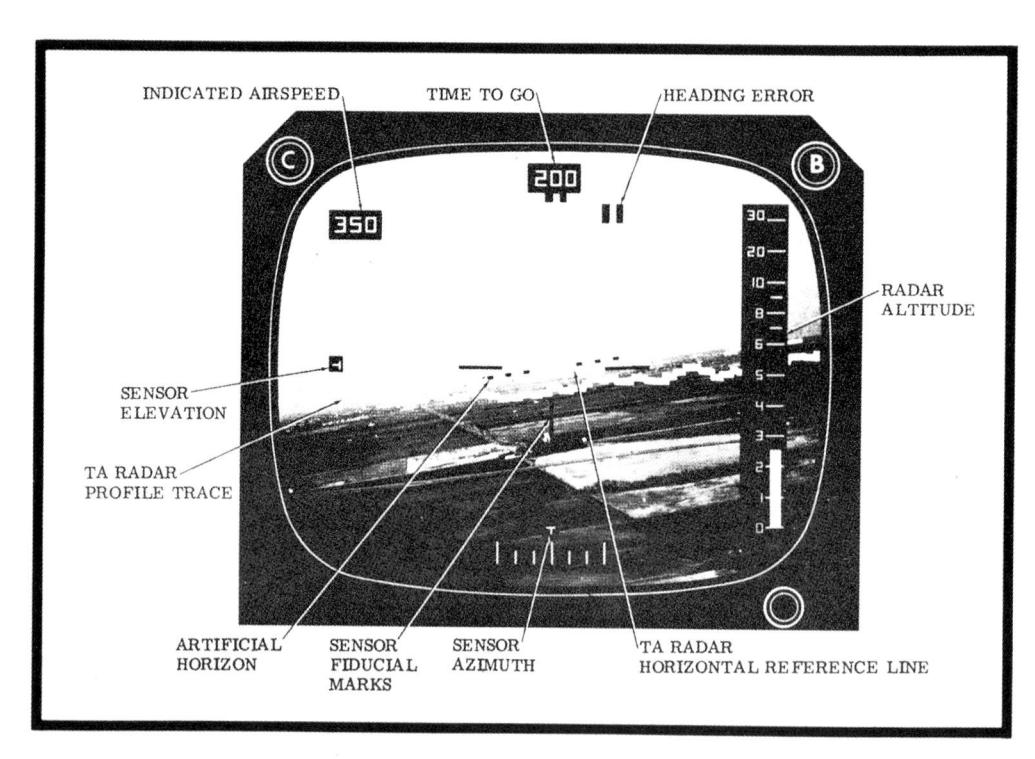

Left: The pilot and co-pilot displays can present either just the television or forward looking infrared sensor picture, with associated flight information as in this illustration, or the terrain avoidance trace can be superimposed if desired. The crews feel much more comfortable with the visual backup to the radar presentation. (Boeing)

Right: An EVS equipped B-52G from the 416th Bomb Wing takes off. (U.S. Air Force).

Below right: The hostile penetration environment into which SAC is targeted presents a variety of challenges—airborne radar, ground radar, SAM sites, anti-aircraft gun sites, and airborne interceptors. (Boeing)

infrared (FLIR) and the TV picture (although both units make similar representations on the screen, via the computer) an overlayed terrain avoidance profile trace on either EVS or FLIR, and alpha-numeric symbology which tells them their radar altitude, time to go before bomb drop, indicated airspeed, heading error/bank steering, artificial horizon, attitude and the position of the sensor.

Despite the complexity, the display is very cleverly done, and is easily understood by the crew.

The navigators, similarly, have a choice of selecting either FLIR or TV, and have the same symbology as the pilots except for the artificial horizon indication.

John See, who helped shepherd the programme through its long gestation period, recalls that the programme had not received all the backing it might have until General Bruce Holloway, CINCSAC, test flew it and enthusiastically endorsed it. From that point on, the programme had clear sailing.

Crews took to EVS instantly, in contrast to their guarded and totally understandable reluctance in regard to the original terrain avoidance equipment. EVS improved safety significantly, because it provided an extremely reassuring cross check with the terrain avoidance radar. In addition, obstructions like television antennae would show up on the television set, but not on the terrain avoidance radar. The presentation of flight information simplified the pilot's instrument cross check, and as an incidental benefit, actually reduced instrument panel space requirements, always at a premium in an ever modified bomber.

A special environmental control had to be built into the system to provide heat and air movement; anti-icing had to be provided as well as a special window washing system to keep the FLIR and TV camera windows clean.

The entire programme, which cost $248·5 million, provided the G and H models with a totally new capability, one which added both striking power and safety.

It is illustrative that this successful programme, which had adequate backing and funding, took 11 years between initial concept and full fleet installation.

Black Boxes, Defensive and Offensive

Electronic Counter Measures (ECM) have come a long way since the primitive jamming done by noise generators or strips of tin foil in World War II. The B-52 was designed from the start with receiving sensors, powered jammer transmitters and chaff dispensers. The latest modifications have vastly increased the size and scope of ECM equipment, more than 6,000 pounds of which are currently installed in each aircraft.

The reason for this increase is the growth of sophisticated threats including ground radar systems, airborne interceptor systems and a vast network of surface-to-air missiles. Only the huge size of the B-52 made it possible to add the required gear to offset Soviet advances.

B-52 ECM evolved over the years through a series of phased improvements to enhance its penetration ability. "Threat radars"—those used by the enemy—operate across a frequency spectrum ranging from A-Band (0 to 250 megahertz) to J-Band (10,000 to 20,000 megahertz). Early warning (EW) and ground controlled interceptor (GCI) radars operate in the low frequency bands, and their purpose is to provide long range detection of incoming bombers and vector interceptors for engagement. Surface-to-air missile (SAM) radars and their guidance links generally operate in the mid-frequency bands. SAMs include stationary and mobile systems that can be deployed to support tactical operations. Airborne Interceptor (AI) radars operate in upper

AWACS

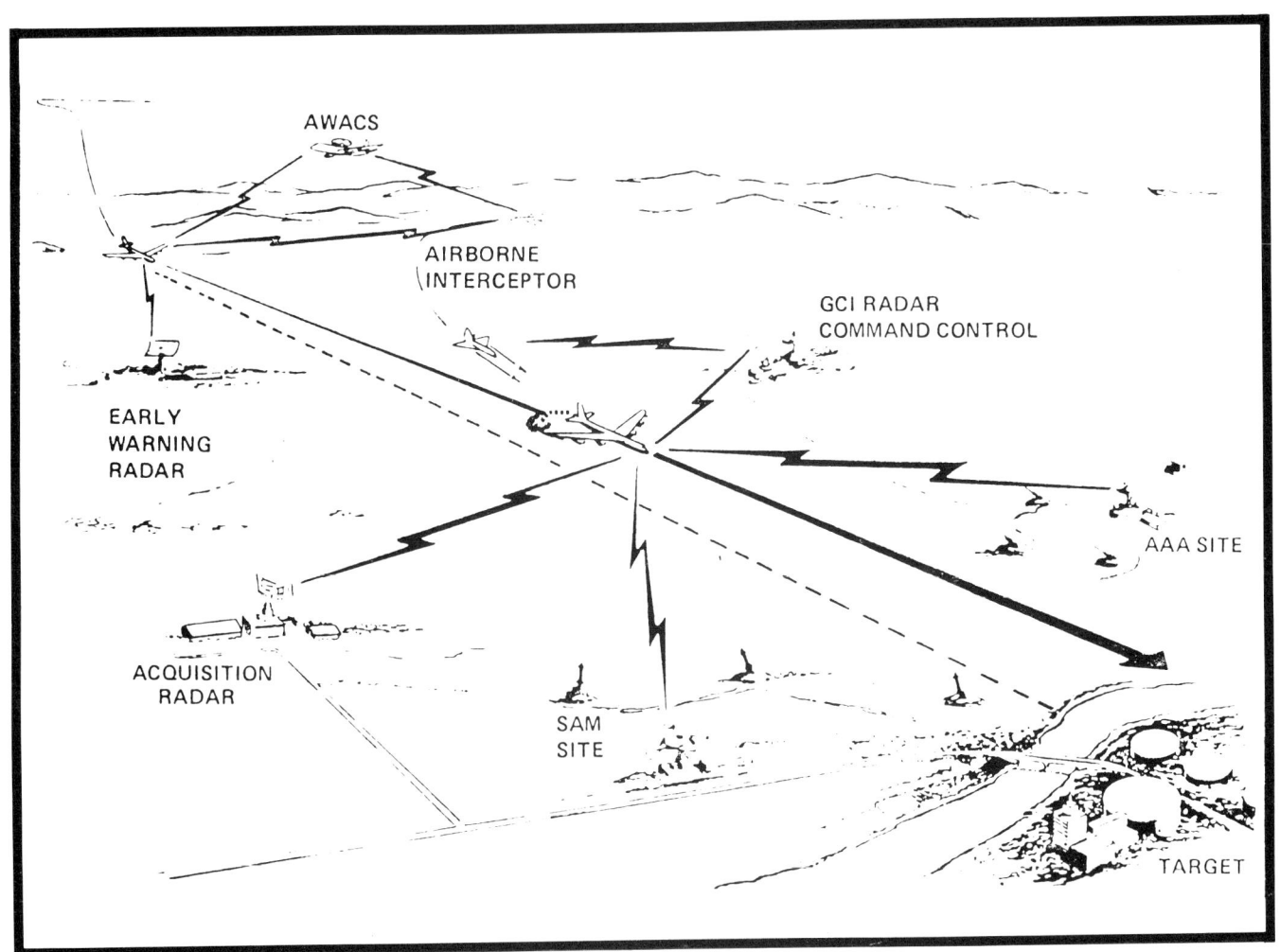

AIRBORNE INTERCEPTOR

GCI RADAR COMMAND CONTROL

EARLY WARNING RADAR

AAA SITE

ACQUISITION RADAR

SAM SITE

TARGET

Sensors

	ALR-46 Warning Receiver			
ALR-20 Panoramic Receiver			ALQ-153	
ALQ-122		ALQ-117		
	ALQ-155 Receiver			

Jammers

ALT-32	ALT-16	ALQ-117	
	ALQ-155 Trans.		

Expendables

ALE-24 Chaff	ALE-20 Flares

Frequency →

	56	57	58	59	60	61	62	63	64	65	66	67	68	69	70-84	85-
	0			I			II/III			IV			V			VI+	Potential Follow-on
Sensors	APR-9 APR-14 APS-54			APR-9 APR-14 APS-54			APR-9 APR-14 APS-54 ALR-18			ALR-20 APR-25 ALR-18			ALR-20 APR-25 ALR-18			ALR-20 ALR-46 ALQ-117 ALQ-122 ALQ-153 ALQ-155	• Integrated defense system – Advanced receivers – Central processors
Jammers	14 ALT-6B			10 ALT-6B 2 ALT-13 1 ALT-15H 1 ALT-16			5 ALT-6B 2 ALT-13 2 ALT-15H 1 ALT-15L 1 ALT-16			5 ALT-6B 2 ALT-13 2 ALT-15H 1 ALT-15L 1 ALT-16			4 ALT-6B 6 ALT-28 2 ALT-32H 1 ALT-32L 2 ALT-16			10 ALT-28 2 ALT-32H 1 ALT-32L 2 ALT-16 1 ALQ-122 4 ALQ-117	• ALQ-117 Update • Monopulse • Comm/D-L/IFF • EOCM Jammers
Chaff Flares	2 ALE-1			2 ALE-1			8 ALE-24 6 ALE-20			8 ALE-24 2 ALE-25 6 ALE-20			8 ALE-24 2 ALE-25 6 ALE-20			8 ALE-24 12 ALE-20	• Expendable Jammers

Current Capability: B-52D | B-52G/H

Future Capability: B-52G/H

Left: A sanitized (i.e. critical frequencies removed) chart which shows the general array against which certain ECM equipment is programmed. Material like this is difficult to present, because while a chart may in itself be unclassified, it may become so in association with another chart. (Boeing)

Below left: The total array of B-52 ECM equipment as it has been improved over the years. Only the great size of the aircraft permitted this increase in space and weight. (Boeing)

frequency bands to enable use of the small antennae consistent with the limited space in the aircraft.

The phased improvements, many of them highly classified, are shown in the accompanying table. They have been progressively updated to increase the effective radiated power, expand the frequency coverage, and refine the countermeasure technique needed to counter enemy threat improvements.

Multiple threats are now countered through the use of a high speed computer which phases the power of the onboard equipment exactly as needed to match to time and direction of enemy signals.

Present day systems have the capability to detect the presence of threat radars, determine the appropriate jamming technique, and automatically programme the jammers against a variety of enemy targets, with minimal operator participation. The Electronic Warfare Officer (EWO) has become a systems monitor/supervisor, who can concentrate his attention on a limited number of controls to inhibit or enable the operation of selected jammers.

During the "11 day war" over Hanoi and Haiphong, EWOs came into their own. Some operations analysis personnel had predicted that the B-52 would be unable to survive in the Hanoi airspace. Unable to manoeuvre to avoid the SAMs, as the F-4s and F-105s could, the B-52s were supposed to be easy targets. In the actual event, the concentrated ECM capability of the B-52s, supplemented by other forces, proved to be adequate against what has been described as the most sophisticated, most concentrated anti-aircraft defence in the history of the world.

As noted before, pilots were often bemused by the intercom chatter of the EWOs on a raid. The pilots, who could see the telephone pole-like SAMs coming up, often 20 at a time, were fully aware of the danger. The EWOs often took genuine delight in each "Uplink" or "Downlink" signal, as if they were bird watchers sighting a new species. There is a chess like quality about the ECM game that seems to attract a certain type of individual to the field. Most B-52 pilots say that they can spot an EWO at 50 feet in a crowded room, and they say it with affection and respect.

The ECM equipment of the B-52 is highly sensitive, and the accompanying charts are as far as one can go without divulging classified material. B-52Ds have been equipped to the Phase V level illustrated; B-52Gs and Hs go to Phase VI, and there are further improvements scheduled for the next decade.

The Phase VI equipment was needed desperately to penetrate modern Russian defences. The modification included addition of the AN/ALQ-117 countermeasures set, additional AN/ALT-28 countermeasures transmitters and AN/ALE-20 flare dispensers. The AN/ALR-46 warning receiver and the AN/ALQ-122 SNOE (Smart Noise Onboard Equipment) brought the total cost of Phase VI up to $362·5 million.

Offensive Avionics Systems

The offensive equipment of the B-52Gs and Hs suffered the same sort of obsolesence as did that of the B-52Ds—there was, after all, only seven years between the delivery of the first D model and the last H model. Few things have changed faster than the electronics industry in the last ten years, and there was great potential for improvement.

The Air Force recognized this in the early 1970s, when it became apparent that the B-52 was going to be around much longer than anyone had anticipated, and in 1975 authorized some major studies to conduct a Long Life Cycle Avionics study which was to stimulate a number of possible replacements for existing equipment in the B-52s, and which would determine what the most effective mix would be over the new extended life of the aircraft.

The results of the study formed the basis for the analysis of airborne computer controlled electronic systems to fulfill SAC's near term needs for cost effective improvements to the B-52 G and H offensive avionics. The desire was to achieve the most effective combination of improved performance, lowered life cycle costs and compatability with equipment retained in the aircraft. The new computational sub-system arrived at full redundancy via new controls and displays, a modified advance capability radar, a complex computational system consisting of two avionics processors and four data transfer units; a weapons control and delivery system, with three missile interface units, a new radar altimeter, a doppler velocity sensor, two inertial navigation sets, and an attitude heading reference set.

This intricate and precise equipment was combined into the Offensive Avionics System (OAS) to include the prime mission equipment (ie the radar navigation and bombing systems), the associated hardware and software, training, support equipment and spares. The OAS had to be versatile enough to handle gravity weapons, short range attack missiles (SRAM) and air launched cruise missiles (ALCM).

Many of SAC's requirements were complementary; the vastly improved navigational capability provided a continuous pinpointing of the aircraft's position, which made both low level penetration and weapons delivery more precise.

The OAS was first flown on September 3, 1980, by a Boeing crew in a B-52G at Wichita. The complex OAS features gain their capability from sophisticated electronics and the human interface which is used to ensure that the black boxes receive inflight verification from known check points. The actual process is extraordinarily demanding at first, and will require intensive crew training.

The OAS programme, with its $1·662 billion cost, will be retrofitted to all G and H aircraft by 1987.

Another modification, already in half of the G and H fleet, will complement the OAS with the ultra modern Air Force Satellite Communications System. This will provide instant high priority communication world wide for operational control and command of forces. The retrofit modification includes the primary controls, the printer and keyboard, a new antenna and the receiver transmitter and modem (modulator and demodulator). Cost is expected to be $108·7 million, and the B-52s will have an ensured reporting and recall capability at any point on the globe.

This almost deadening recitation of modifications to the B-52 are but a part of the total it has received. The aircraft has literally been transformed beneath its skin, always retaining a similar external shape, but virtually remanufactured to new standards of excellence time and again. The modifications wedded to the epochal original airframe design have permitted the B-52 to be in service longer than any other first line major weapon system in history.

There may be other changes still; there is real thought being given to re-engining the aircraft with modern jumbo-jet type engines, and who can say that yet another missile will not be devised to operate from the ageing sub-sonic warrior. Perhaps the really big story will be when the first grandfather-father-son-crew flies a B-52.

Above: *The "Rivet Ace" aircraft on which the advanced ECM equipment was tested in a spectacular photograph which catches a phenomenon caused by certain atmospheric conditions. The water vapor temperature is lowered as it flows around the aircraft, increasing the relative humidity and causing water droplets to form in a localized "fog". The phenomenon is rare, and almost never caught on film.* (Boeing)

Above right: *The latest programmed defensive avionic systems programmed for the B-52 G and H.* (Boeing)

Right: *Just as the defensive electronics have had to be improved, so have the offensive systems. The B-52Gs and Hs are in the process of receiving a complete updating of their offensive systems, to include integration of SRAM and ALCM capability. It is "beneath the skin" modifications like this which extend the B-52's service life.* (Boeing)

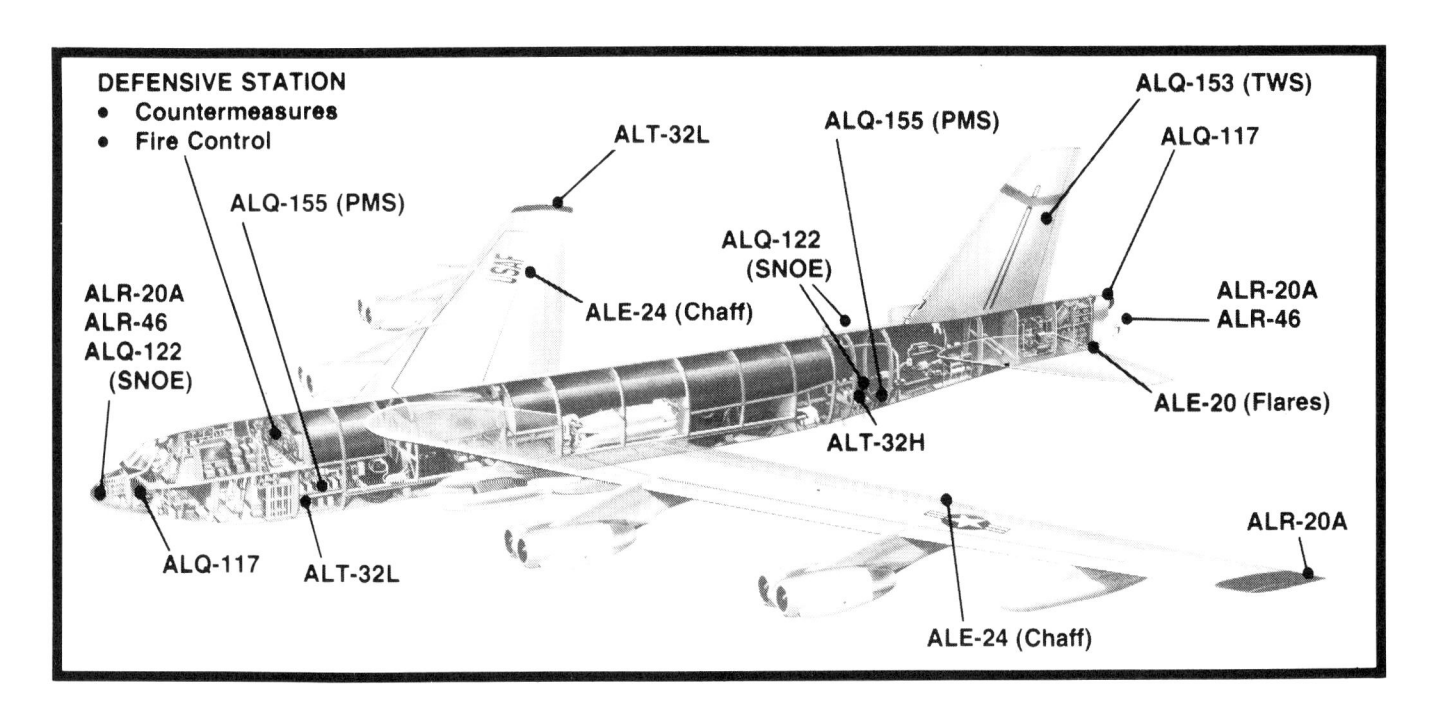

DEFENSIVE STATION
- Countermeasures
- Fire Control

- ALT-32L
- ALQ-155 (PMS)
- ALQ-153 (TWS)
- ALQ-117
- ALQ-155 (PMS)
- ALQ-122 (SNOE)
- ALR-20A
- ALR-46
- ALQ-122 (SNOE)
- ALE-24 (Chaff)
- ALR-20A
- ALR-46
- ALE-20 (Flares)
- ALT-32H
- ALQ-117
- ALT-32L
- ALR-20A
- ALE-24 (Chaff)

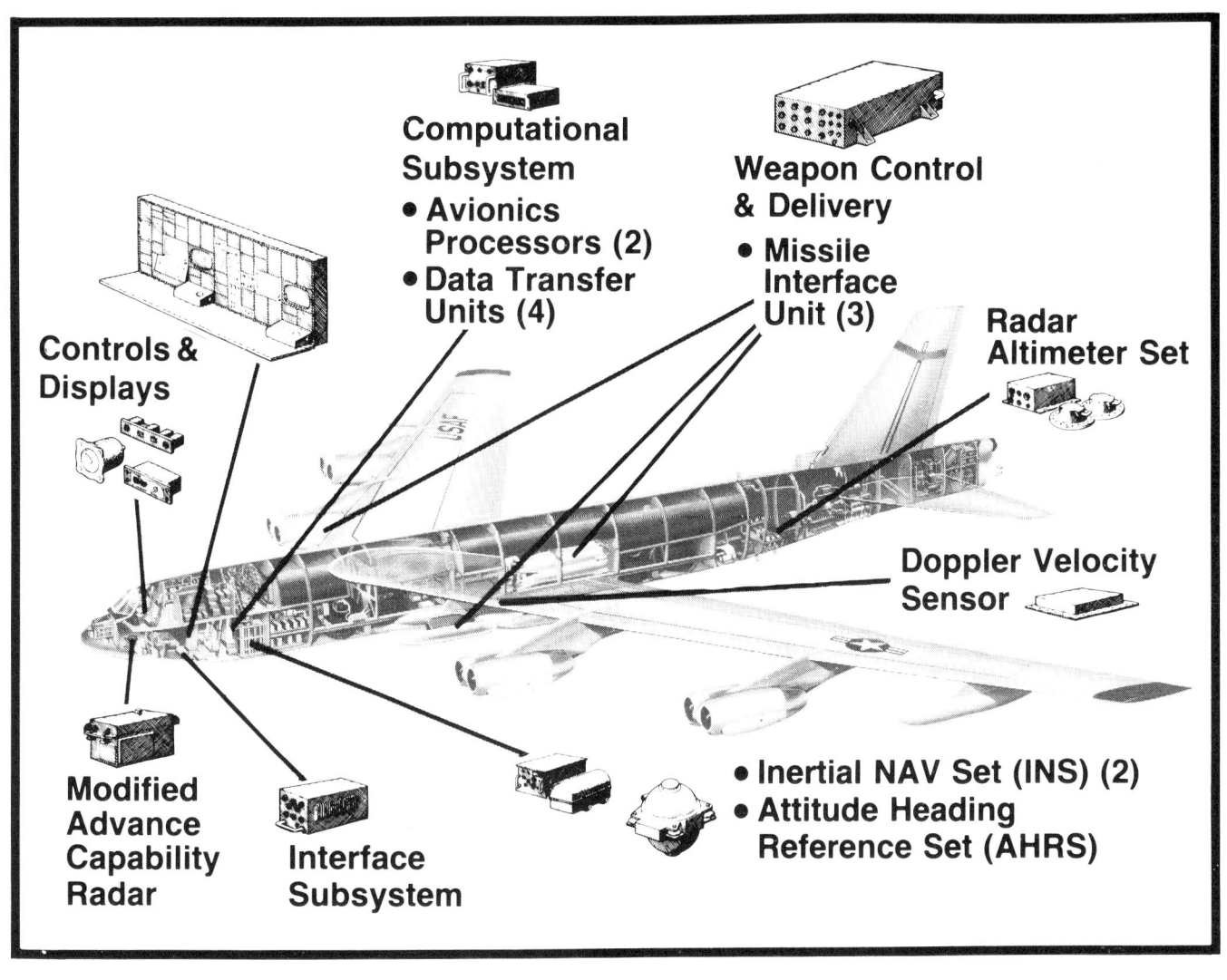

Computational Subsystem
- Avionics Processors (2)
- Data Transfer Units (4)

Weapon Control & Delivery
- Missile Interface Unit (3)

Controls & Displays

Radar Altimeter Set

Doppler Velocity Sensor

Modified Advance Capability Radar

Interface Subsystem

- Inertial NAV Set (INS) (2)
- Attitude Heading Reference Set (AHRS)

Left: *On September 4, 1980, the first B-52G completely equipped with EVS and the offensive avionics system (OAS) takes off from Wichita.* (Boeing)

Below: *The introduction of new equipment compounds the always critical problem of "nuclear hardening", the protection of systems from the effect of radiation in a nuclear environment. Here a B-52, completely instrumented and equipped with modern systems, has been laboriously tugged up a pyramid-building like trestle system to undergo a test for hardening.* (Boeing)

MISSILES, BOMBS AND GUNS

The B-52 was envisioned from the very first to have a multiple mission capability, although its primary purpose was to drop gravity fall nuclear weapons. Still, there had been provisions made for conventional bombs, and there had even been talk of fitting that worst of all air launched missiles, the Bell Rascal, early in the programme. Fortunately, the complete and abysmal failure of the test programmes made on the B-47 removed the Rascal from consideration.

The first missile used extensively on the B-52 was the GAM-72 Quail, later called the ADM-20 for Air Decoy Missile. The Quail, a diminutive missile produced by McDonnell Douglas had as its sole purpose the confusion and saturation of the enemy's defences by presenting a radar target identical to that of the B-52 from which it was launched. The missile was only 0·3% of the weight of the bomber, but it had equipment onboard which permitted it to duplicate both the flight characteristics and the radar appearance of the B-52 on enemy scopes.

The Quail was preprogrammed to fly a flight path which simulated the B-52 penetration manoeuvres, maintaining the same speeds and altitudes.

The tiny Quail was 13 feet long, 3 feet high, and had a 5 foot 6 inch wing span. The wings folded to fit in the bomb bay,

Below: *The GAM 72 Quail missile was capable of portraying a B-52-like image and flight path on enemy radar screens. Fully deployed the tiny aircraft had a thirteen foot length, five and one half foot wing spread and a speed range from Mach .6 to .9. (Boeing)*

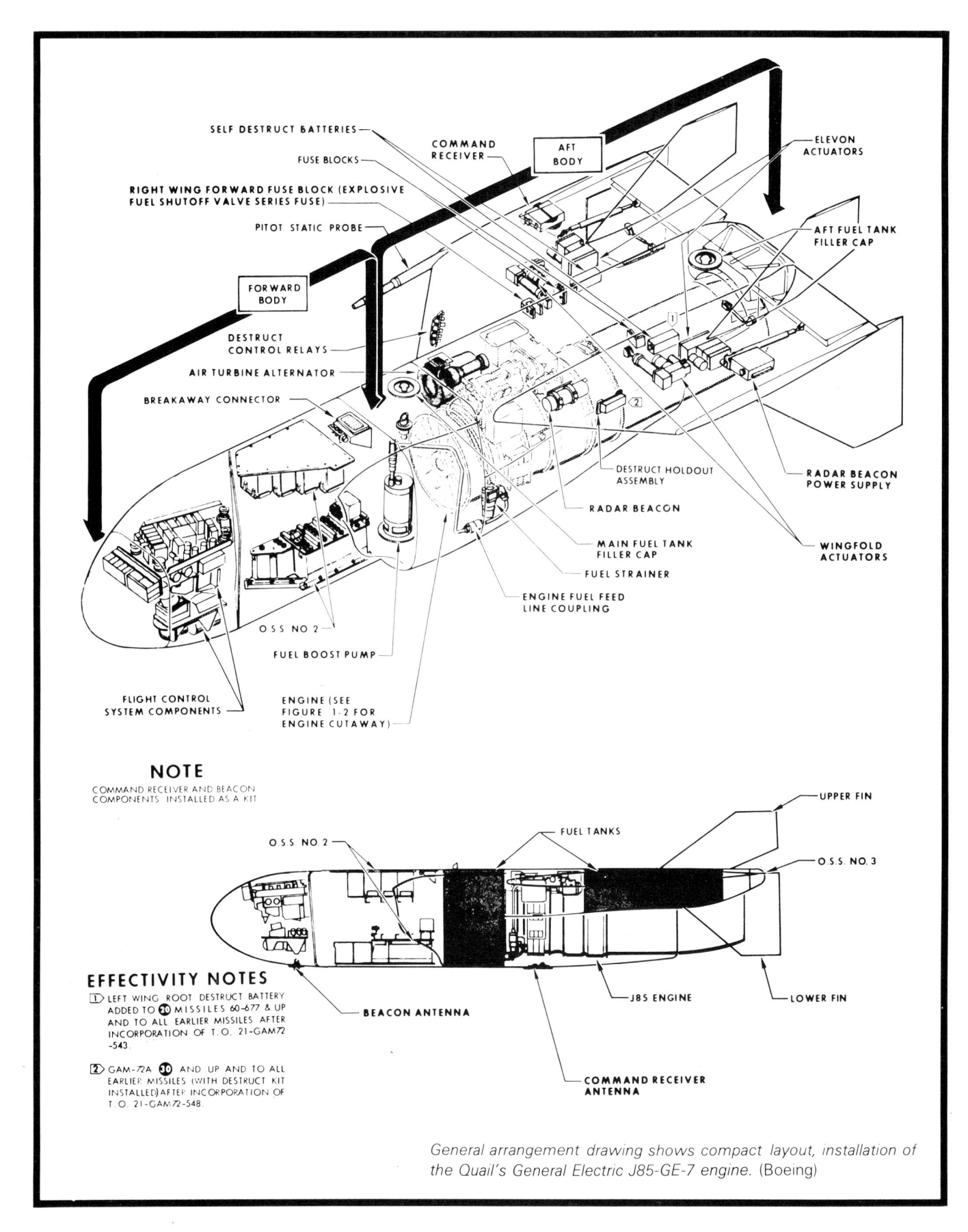

General arrangement drawing shows compact layout, installation of the Quail's General Electric J85-GE-7 engine. (Boeing)

reducing in width to only 2 feet 4 inches. Most of the forward half of the missile was a radome housing the flight control system electronics and the radar reflectors. The remainder of the missile consisted of a modified delta wing, fixed fins, fuel tanks and the General Electric J85 engine which had an infrared burner to increase its detectability. Like later missiles, the Quail had a semi-automatic checkout equipment which permitted storing it in an almost ready to launch condition. The flight programme was preset on the ground but could be changed in flight by the radar navigator.

To launch, the radar navigator opened the bomb bay doors, and selected a missile by flipping a toggle switch. The launch gear tracks lowered the missile into the airstream, and the wings extended and the engine started automatically. Once launched the Quail was independent of the bomber. Speed range was from Mach ·6 to ·9, and the range varied from 400 miles at high altitude to only 39 at low altitude.

First flown in 1958, the Quail entered service in 1960, and served until 1978 when changed operational requirements dictated its retirement. The B-52 could carry four Quails in addition to its normal bomb load, and it may be said in truth that a Quail never fooled an enemy in anger.

Above: *SAC crews liked the Hound Dog, for it not only provided "insurance" during an actual penetration, it gave extra power for takeoff. The Hound Dog tank could be topped up in flight from a B-52.* (U.S. Air Force)

Right: *Unusual "bottom view" shows 42 foot long Hound Dog nestled in its pylon. The missile could carry a four megaton load.* (Air Force)

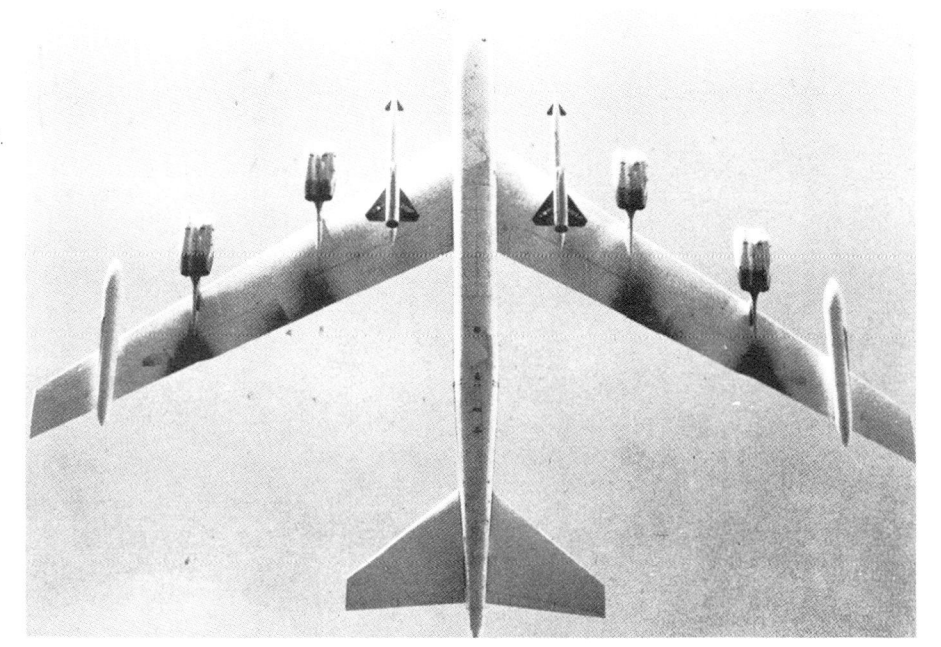

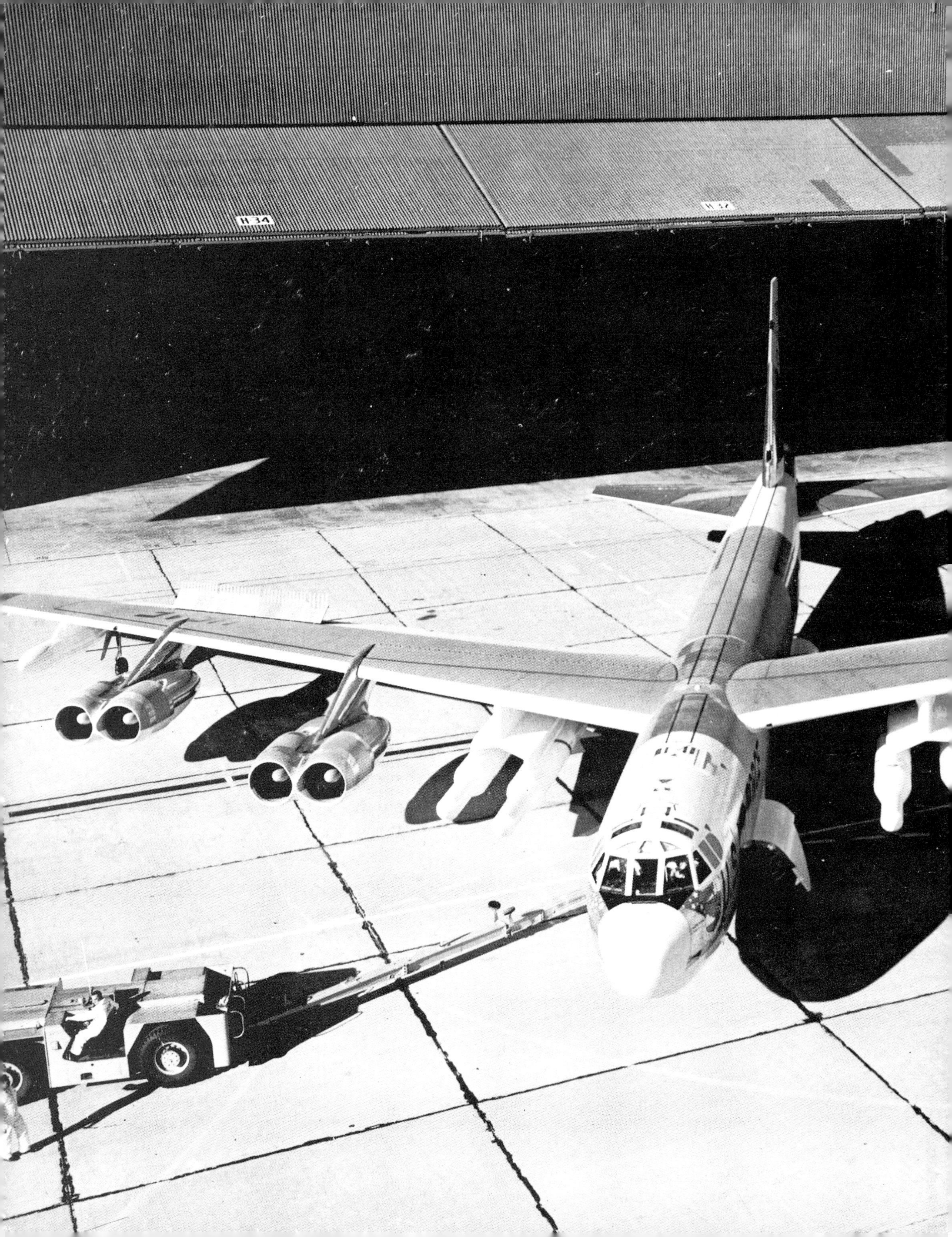
1134

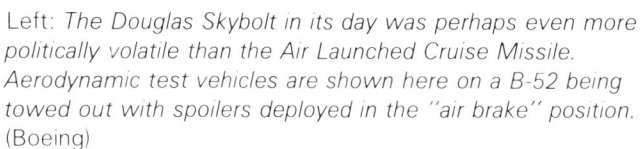

Left: *The Douglas Skybolt in its day was perhaps even more politically volatile than the Air Launched Cruise Missile. Aerodynamic test vehicles are shown here on a B-52 being towed out with spoilers deployed in the "air brake" position.* (Boeing)

Above: *Skybolt was an extremely clean looking weapon, with a very stylish pylon mount.* (Boeing)

Hound Dog

The first SAC launch of a Quail had occurred on June 8, 1960, four months *after* the first flight of a much more sophisticated missile, the Hound Dog. The Hound Dog (GAM-77/AGM-28) was a much larger weapon, intended for stand-off use as an air-to-surface attack missile. All B-52G/H aircraft were equipped to carry two Hound Dog missiles on the wing pylons between the fuselage and the inboard engine nacelles. The B-52 could use Hound Dog's Pratt & Whitney J52-P-3 turbojet engines for extra power on heavyweight take-offs.

Built by North American Aviation (now Rockwell International) the Hound Dog made its first flight in 1959, and was operational in SAC from 1961 to 1976. At its peak, there were 593 Hound Dogs in the inventory.

The Hound Dog had an inertial guidance system which was updated by the B-52's onboard system just prior to launch. Range was as much as 700 miles, and speed was as high as Mach 2·1. The flight path could be profiled for tree top level, or for as high as 55,000 feet.

SAC crews were at first worried about the increased weight and complexity of the missile, but soon found out that the increase in available thrust of the J52s more than made up for the weight. The missile's tanks could be topped up by the B-52 in flight.

The inertial navigation system of the Hound Dog could also be used as a back-up to the B-52's, and at least one crew overcame an emergency by doing so. It was immune to jamming, and like the Quail, could have its mission profile changed before launch by the radar navigator.

While the Quail was purely a decoy, the Hound Dog not only distracted the radar system, but was capable of delivering a one megaton warhead on target. It was a first class cruise missile before the term became popular—or controversial. Some models were fitted with a terrain following system which increased its chances of penetration.

Skybolt

While the Hound Dog and the Quail were successful and became fully operational, they are probably less well known to the world in general than a missile which had a career studded with misfortune, and which ended in a political crisis that came as close as anything has to impairing relations between the United States and Great Britain. It was the Douglas Skybolt, an air launched ballistic missile (ALBM) called the GAM-87A.

This rather handsome weapon had its origin with two other ill fated Air Force projects, the WS 110A, the so called chemical bomber which was to have used exotic fuels, and the even more radical WS-1252 nuclear powered bomber. Both of these had General Operational Requirements which included an advanced air-to-surface missile. In its subsequent history, the Skybolt went through five separate sets of requirements, three demonstration programmes and no less than 37 proposals for manufacture.

Built by the Douglas Aircraft Company, in concert with a host of other contractors including Boeing, Aerojet General, Nortronics and Bendix, the two stage inertially guided missile was to have a hypersonic speed and a range of over 1,000 miles.

The weapon offered a number of tactical advantages; it greatly multiplied the problems of the Soviet defence, because it could be launched from any point on the globe, and it could be used as a bruising method of rolling back defences to provide a safe corridor for its carrier aircraft to penetrate with heavier gravity bombs.

An agreement had originally been made between the US and Great Britain that the Skybolt was to be a joint use weapon, replacing the RAF's Blue Streak missile. The B-52H was to carry four Skybolts, while the Avro Vulcan Mark II was to carry two. The Skybolt was seen by many in Great Britain as the best guarantee of maintaining an independent nuclear deterrent.

Douglas won the contract to develop the weapon on May 26, 1959; shortly thereafter it awarded development con-

Right: *Launch and ignition of Skybolt. A completely successful test took place on day of cancellation. Air Force Program Officer Brigadier General David "Davey" Jones had held a press conference detailing success, which appeared on the same page of newspapers in which President Kennedy announced that Skybolt "was not technically within our grasp". Great embarrassment on all sides!* (Douglas)

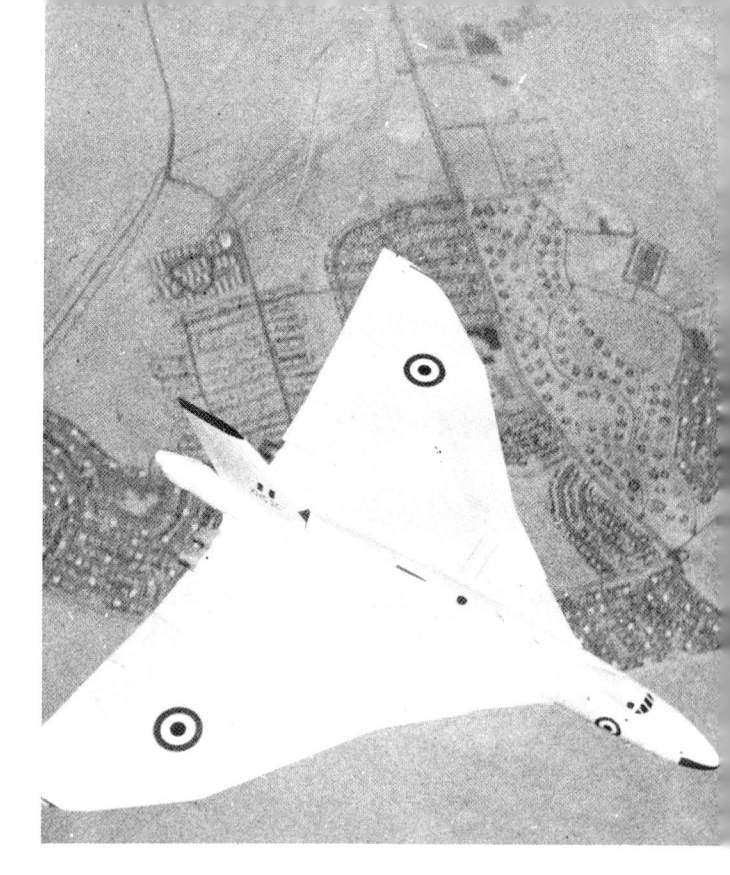

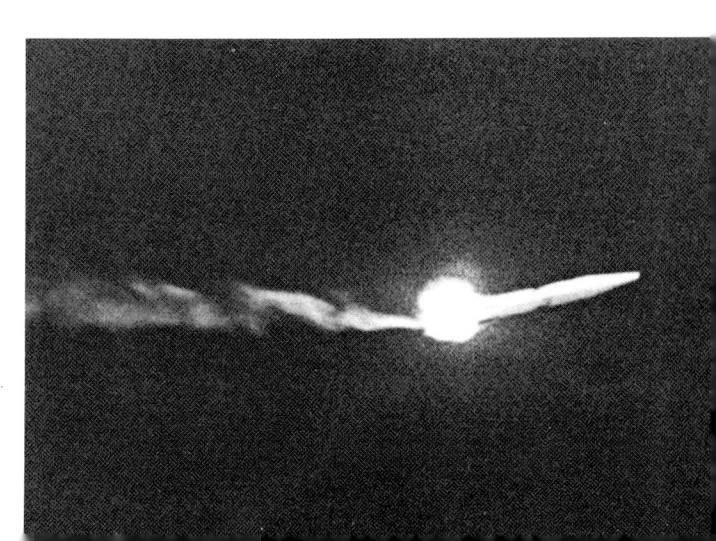

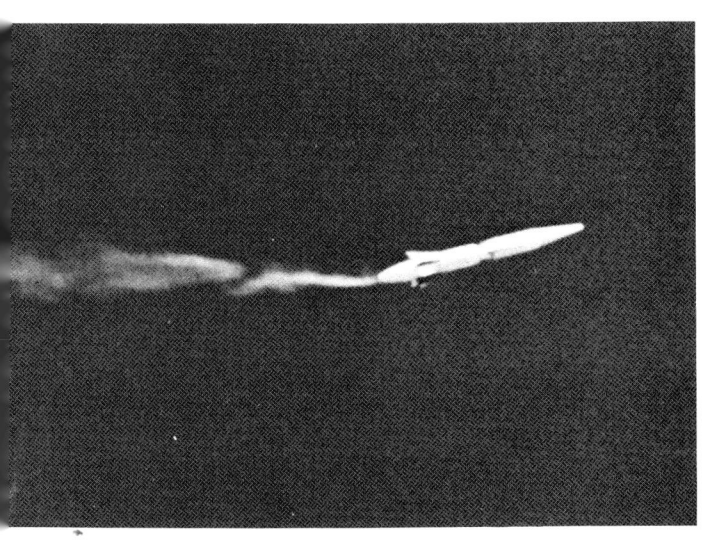

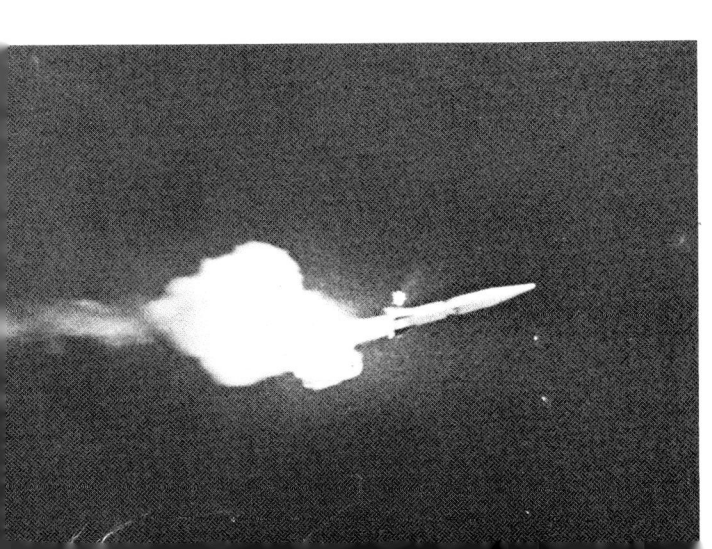

Left: *Great Britain planned to arm Avro Vulcan's with two Skybolts and the programme cancellation was a bitter blow to the RAF. The companion B-52H would have carried four.* (U.S. Air Force)

tracts to Aerojet General for the propulsion system, to General Electric for the re-entry vehicle and to Nortronics for the guidance system. Boeing and Avro assisted the firm to ensure compatibility with the B-52 and the Vulcan.

The programme was unpopular in the press in both England and the United States, for different reasons. The US objected to the fact that the USAF was picking up all development costs, and that England was paying only for production costs of the weapons it accepted. Great Britain was upset that even that amount of money should go to the US.

The Skybolt followed a checkered test pattern, and like all similar projects, costs skyrocketed over the life of the programme. Secretary of Defense Robert McNamara was against the programme and eventually the recommendation was made to cancel the whole thing on technical grounds, although the spectre of reduced cost effectiveness had begun to haunt the programme as the Minuteman missile became operational.

Britain was offered the Polaris missile as a substitute for Skybolt in the event that the air launched weapon was cancelled. This too met with a mixed reaction, in part because the permission to base US Polaris equipped submarines in Scotland had been a part of the original Skybolt agreement between the two countries.

There followed one of those cosmic comic coincidences which cause Program Directors to go grey overnight. On the very day that President Kennedy met Prime Minister Harold Macmillan in Bermuda, and formally cancelled the programme on the grounds that Skybolt "was not technically within our grasp", a totally successful Skybolt launch was made. The success of the test so elated Program Director and former Tokyo raider Brigadier General David M. "Davey" Jones that he held a formal press conference, detailing the 100% perfect flight.

Somewhat to his and the President's embarrassment, the stories of both the "cancellation for technical reasons" and the perfect flight appeared simultaneously on the front pages. Jones was summoned to Washington, dressed down, and from then on the word went out that no Program Director could have a press conference unless the material was cleared at the top in the Pentagon.

In the end, Great Britain reluctantly accepted the Polaris, and the Skybolt programme at long last died.

SRAM—The Untouted Winner

The very great increase in Soviet air defence capability in the 1960s forced a revision of USAF thinking about the penetration capabilities of the B-52. It was apparent that to reach its targets, no matter what altitude was flown, nor what on-board equipment was available, the bombers would literally have to blast their way in, destroying radar and SAM sites as they went. Multiple weapons of relatively low yield seemed to be the answer, and the AGM-69A, SRAM for Short

Range Attack Missile, was developed for both the B-52 and for the General Dynamics FB-111.

The SRAM has a nuclear punch equivalent to a Minuteman III warhead, with similar accuracy, and once launched is virtually impervious to either electronic countermeasures or interception. Fourteen feet long, and weighing about 2,200 pounds, the missile has a range at high altitudes in excess of 100 miles; this comes down to about 30 miles at low altitudes. Its small radar cross section makes it very difficult to pick up on radar, and its hypersonic speed makes interception impossible during its average three minute flight from bomber to target.

The missile can be operated in a semi-ballistic mode, where it is launched and flies an arcing trajectory, much like an artillery shell, to the target. The rocket motor furnishes an initial launch pulse, and a lower rate sustaining pulse while in flight. The missile coasts to the target when fuel is exhausted.

A low level flight mode is made possible by the use of an accurate radar altimeter, which allows the SRAM to hug the terrain. The missile can also combine semi-ballistic and low level modes, if the nature of the terrain requires it.

The B-52 can carry up to 20 SRAMs, six on each wing pylon and as many as eight in the bomb bay mounted rotary launcher, although usually only six are carried. SRAMs can be launched from the launcher or the pylon at the rate of one every five seconds.

Above left: *The need to strike many more targets on the way in on a penetration brought about the smaller yield SRAM, a supersonic air-to-ground missile. The B-52 can carry twenty, eight internally and six on each pylon.* (Boeing)

Left: *SRAM has a nuclear punch equivalent to the Minuteman III warhead, with similar accuracy. Once launched it is virtually impervious to either electronic countermeasures or interception. Test launch shown here.* (Boeing)

Below: *B-52G with full load of SRAMs taxies out.* (Boeing)

Top: *"Rivet Ace" test aircraft with SRAMs being tested for compatibility.* (Boeing)

Above: *Close-up of pylon installation.* (Boeing)

Right: *The SRAM internal rotary launcher, which can be fired at rate of one missile every five seconds.* (Boeing)

The most amazing thing about the SRAM is the very little attention that was paid to it during its development, acquisition and deployment. Unlike the neutron bomb, the B-1, or the Air Launched Cruise Missile, the SRAM did not attract controversy either in the United States or abroad. It may be that its very name "Short Range Attack Missile" implied a certain lack of threat, and consequently did not make good copy for the press.

Boeing won the contract for development of SRAM on October 31, 1966, and as is the practice in the industry went to major firms to sub-contract important elements of the work. While Lockheed had created the original propulsion system, Thiokol was awarded the contract for the two stage XSR-75-LP-1 solid fuel propulsion system, perhaps the most distinctive element of the missile. Singer's Kearfott Division designed and built the guidance system. Unidynamics was

Top and above: *Head-on comparison of winning Boeing AGM 86B and the losing General Dynamics AGM-109 "Tomahawk".* (Boeing)

responsible for the very critical safe/arm/fuse sub-system, while the inertial guidance gear was supplied by Litton Industries. Stewart-Warner manufactured the terrain clearance sensor, and Delco Electronics provided the missile computer.

After extensive testing, the first live flight was made on July 29, 1969, and on September 24, 1970, the first all-SAC launch was made. The missile became operational on August 4, 1972 with the 42nd Bomb Wing at Loring AFB, Maine.

The production of the missile is just part of the total effort involved, of course, for Air Force personnel have to be trained to conduct missile testing, do maintenance, and mission preparation for flight. The missile apparently has encountered no major difficulties during its operational career, and has apparently met all accuracy and range requirements during proficiency tests. Ultimately 1,500 SRAMs were delivered, and if the replacement for the B-52 or B-1 is ever ordered, additional SRAMs may well be procured.

The ALCM

Perhaps the biggest contractual plum in years, the ALCM (Air Launched Cruise Missile) has had just the opposite sort of publicity as did the SRAM. It has generated intense interest at every level from the daily newspapers to the SALT II talks, and was the subject of one of the most intensive, hardest fought competitions in history. The furore over the cruise missile concept is a little hard to understand, as the basic idea dates back to the Kettering "Bug" of World War I, and the much more famous German V-1 "Buzz Bomb" of World War II. The United States has developed and deployed a number of weapons which could easily be called cruise missiles, ranging from the submarine based Regulus of 1951 through the Matador, Mace, Snark, Bomarc, Talos, Hound Dog and others. The Russians have built and deployed a similar array of weapons with "cruise missile" characteristics.

The lethality and accuracy of the new ALCM and its ability to avoid detection may account for the fact that it has attracted so much notoriety. Basically a small subsonic aircraft with a length of just over 20 feet, and an extended wing span of 12 feet, the ALCM is powered by an efficient 600 pound static thrust Williams turbofan engine which weighs only 150 pounds. The inertial guidance system is combined with a terrain contour matching unit (TERCOM) that provides the accuracy.

In TERCOM, a radar altimeter surveys surrounding terrain as the missile whips along at tree top level, comparing selected points to those on an electronic "map" stored in the tiny onboard computer. The mini-bomber takes the information, integrates it with the pre-planned route, and can skim over a mountain and down into a valley to avoid detection.

With a range of 1,500 miles the ALCM offers a great stand-off distance capability to the penetrating B-52.

Right: *ALCM has just dropped, wings have not yet deployed.* (Boeing)

Exploded view. ALCM deploys in two seconds, obtains thrust on its Williams 600 pound thrust turbofan engine in ten seconds.

The deployed wing span is 115 inches.
The fin-deployed height is 45.5 inches.
Overall length is 168 inches.
The missile weighs 1,900 pounds.

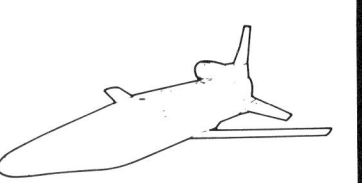

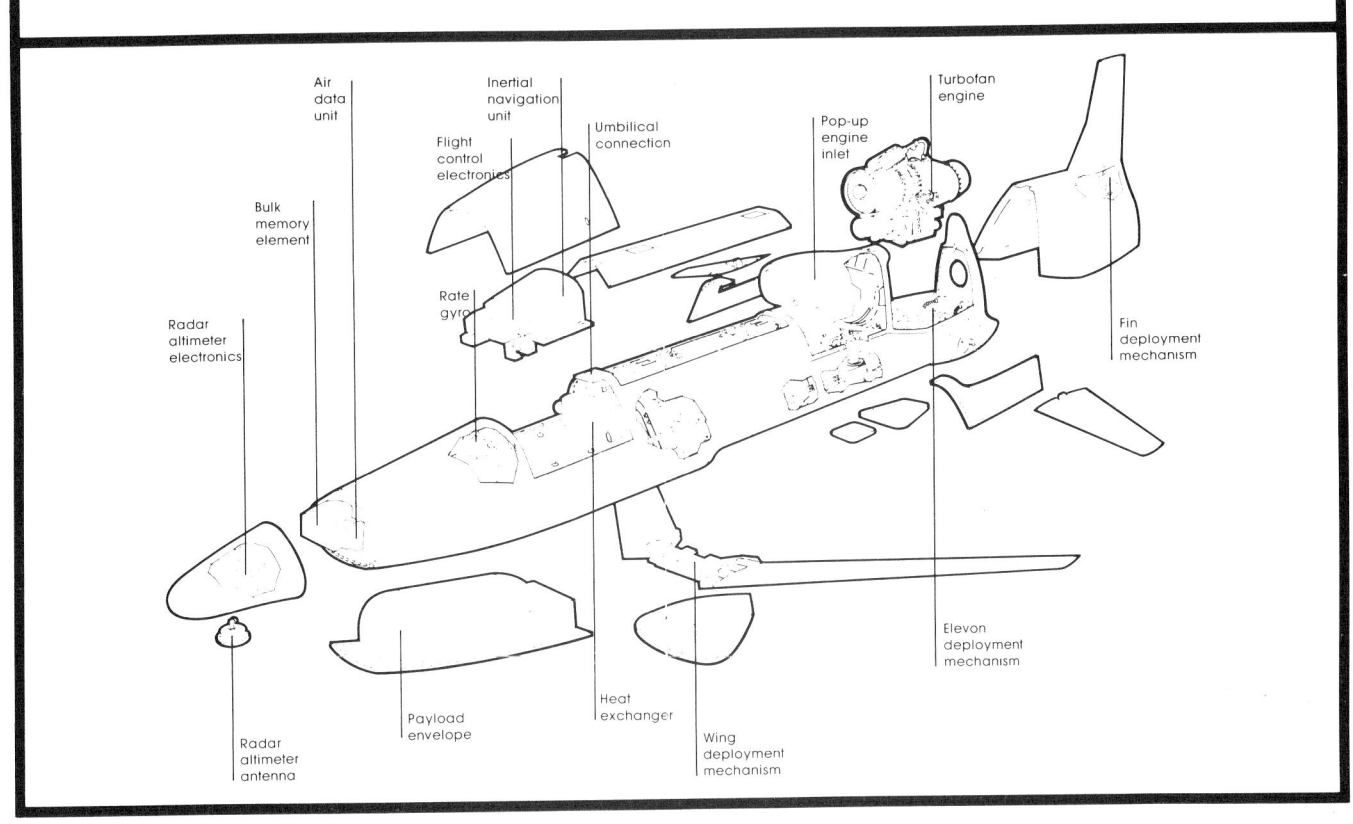

Labels (exploded view):
- Air data unit
- Inertial navigation unit
- Umbilical connection
- Turbofan engine
- Pop-up engine inlet
- Flight control electronics
- Bulk memory element
- Rate gyro
- Radar altimeter electronics
- Fin deployment mechanism
- Radar altimeter antenna
- Payload envelope
- Heat exchanger
- Wing deployment mechanism
- Elevon deployment mechanism

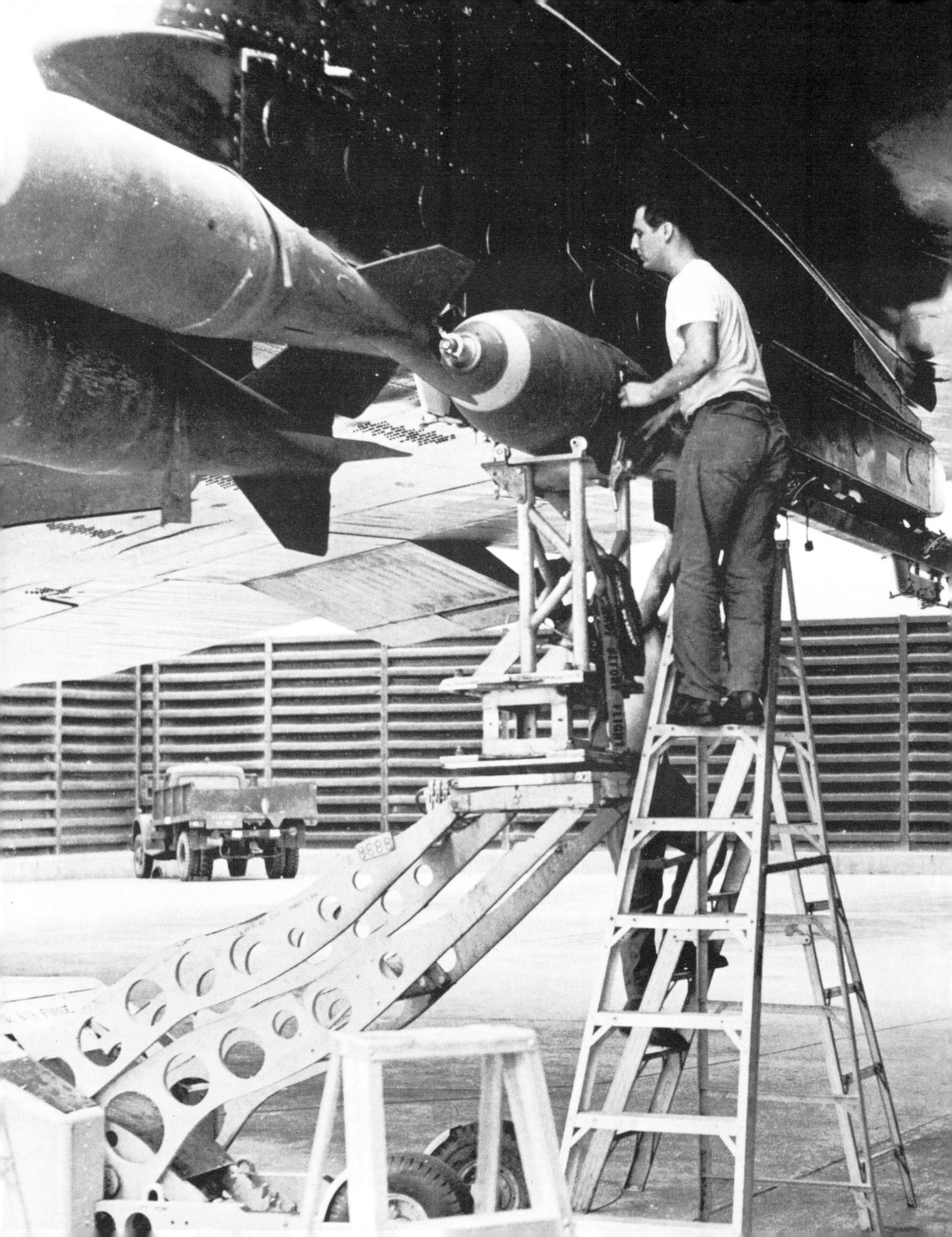

The ALCM actually started life as SCAD, a Subsonic Cruise Armed Decoy, designed to replace the Quail missile. Never built, the missile led to ALCM-A, the AGM-86A which Boeing built when it threw its hat in the cruise missile ring. The ALCM-A was followed by a larger version, the Extended Range Vehicle (ERV) and finally by the larger ALCM-B which bested the General Dynamics Tomahawk (AGM-109) in an intensive fly-off competition.

Boeing was awarded a contract for $141,570,990 on May 2, 1980 to produce an initial 225 missiles. A total of 3,400 missiles will be completed during the production programme. The first unit to get the ALCM will be the 416th Bomb Wing at Griffiss AFB, New York, which has an operationally ready date of December 1982.

Bombs

The B-52 aircraft has a much broader conventional bomb capability than is generally thought, with more than 22 different configurations of bombs and mines being considered standard stores. In addition, special purpose items like the AN/ALE-20 Flare System, the M129E2 leaflet bomb and a whole variety of countermeasure bombs can be fitted.

Left: In over 126,000 sorties, SAC crews dropped more than 2,600,000 tons of bombs, compared to the 955,000 tons dropped by RAF over Europe in World War II. (U.S. Air Force)

Below: Incidental benefit from the "Big Belly" modification was the ability of B-52Ds to carry the GBU-15 modular guided bomb. The bomb uses a TV seeker/data link which provides for target acquisition and lock on before or after launch. There are two configurations, cruciform wing and planar wing. The planar wing is shown here. (U.S. Air Force)

The number of bombs that can be carried is a function of their size, weight and the configuration of the bomb bay. The bombs themselves are usually built up from standard components, which can be arranged in specific combinations in the ammunition build-up area to provide the type desired. For example, the Mk 82 low drag bomb and the Mk 82 Snake eye bomb share the same slender body, but the former has the MAU-93 fin assembly, while the latter has the Mk 15 MOD 1 fin assembly. Parachute packs, release mechanisms, control units, fairings and other items are shared in a similar manner so that the inventory of parts to build up bombs can be significantly reduced, while still providing the Commander a wide option of the bombs to employ.

Gravity Fall Nuclear Weapons

The very existence of a manned bomber as a part of the strategic triad depends upon the use of gravity fall nuclear weapons. This may change in the future, if air defences become inpenetrable, and bombers may cease to exist, or may be just airborne missile launchers.

The accompanying photos show many of the types of bombs that have been used in the past or are in current use. The subject matter is highly classified, of course, but the information shown here has been gathered from non-classified sources.

One of the greatest hazards of any nuclear weapons is that governments and the general public forget what their potential is, and begin discussing them in academic terms. Nothing is more chilling than to see analyses of nuclear war scenarios where terms like "160 US and 120 Russian" really means 160,000,000 US citizens killed and 120,000,000 Russian citizens killed.

The author participated in the last series of live nuclear drops, "Operation Dominic" which took place in 1962. This long series of drops of relatively low yield weapons was designed to provide data for development rather than information on blast effects.

The bombs were parachute retarded, and the drop aircraft would overfly the target and depart on course rather than executing any sort of "breakaway" manoeuvre. After about 40 seconds, there would be a slight bump, as the shock wave of the detonation would hit; then, from behind, would come a gradually increasing ball of light, one which would swiftly spread to the horizon, lighting up the entire Pacific as far as one could see, brighter than daylight. (The crews had become aware after a few drops that shutters and goggles were not in fact absolutely essential when flying away from the blast.)

The entire sky would remain illuminated for what seemed like an eternity, but must have been more than a minute, and then the light would collapse in a curious fashion, like a deflating balloon, coming down back in from the horizon until it was night again.

Our radar navigator, who dropped the weapons, was a crusty old veteran from World War II, not noted for philosophy or profundity, but he had crawled up to the cockpit area to witness the phenomenon, and in a low voice said

Below: *The Mk 17 was the first American air deliverable hydrogen weapon. The first airdrop and detonation test occurred over Bikini atoll during Operation REDWING on May 1, 1956. It is a parachute retarded weapon, and has a yield varying from 10 to 25 megatons.*

Right: *The Mk 28 bomb was carried internally on a number of aircraft, including the B-47 and B-52, and externally on several fighters. The yield can vary considerably, depending upon its configuration.*

"They should make every head of state see this once a year: then they'd know what they were playing with."

And its true; it would be worth the environmental impact to hold a demonstration for heads of state every few years, just to illustrate the holocaustal power at their finger tips.

Guns

The B-47 had been armed with radar directed, remote controlled cannon, so it seemed something of a step backwards when a tail gunner position, in traditional heavy bomber fashion, was planned for the B-52. Four ·50 calibre machine-guns were operated by a Type A-3A fire control system in the B-52A and some B-52Bs, most B-52Cs and a few B-52Ds. The MD-9 system was used on most of the B-52Ds, Es and Fs.

The fire control system has the capability to search, detect, acquire, track and compute the correct lead angle for attacking enemy aircraft, all vital functions in the age of the jet fighter. The gunner is equipped with a periscopic gunsight for manual use should the radar be inoperative.

The gunner uses three basic modes of operation with the system, including search, acquisition and track. The search radar picks up targets out to 8,000 yards, and tracking begins

Below centre: *The huge Mk 36 weighed 17,500 pounds, and could be either parachute retarded or dropped free. Yield was in the megaton range.* (Chuck Hansen)

Below: *The Mk 39 was basically a Mk 15 with a parachute system added. The parachute system was complex, consisting of a 6 foot diameter pilot chute, which triggered the sequential deployment of a 28 foot diameter ribbon drougue chute to stabilize the bomb, a 68 foot diameter octagonal canopy to decellerate the bomb, and finally a 100 foot diameter solid flat canopy which became spherical when deployed. Yield was in the megaton range.* (Chuck Hansen)

at 6,000 yards. Targets of unusually high radar reflectivity can be detected as far out as 20,000 yards.

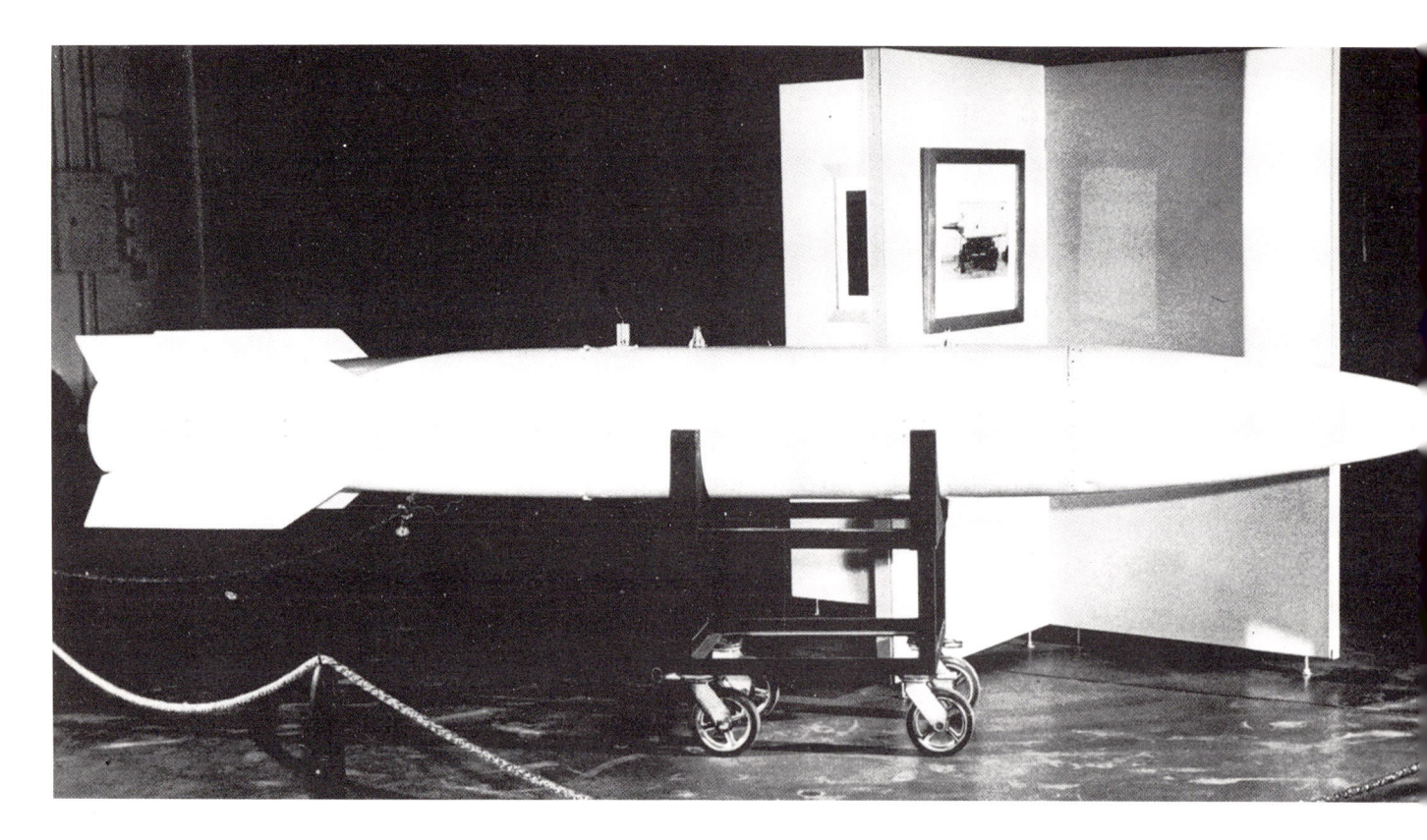

If there are multiple attackers, the gunner can unlock from one, and lock on to another, with a press of a button.

Once in range, the four M-3 machine-guns each have 600 rounds of ammunition to fire.

Thirty-three B-52Bs were fitted with the MD-5 system, which was similar in operation, but used two 20 mm M-24A-1 cannon in place of the machine-guns.

In all models up to the G, the gunner was obliged to jettison the entire turret and make an unassisted bail-out in the event of an emergency. The task was not too difficult unless the aircraft had assumed a nose-down attitude in excess of ten degrees. As the aircraft had a natural tendency to pitch down once the turret was jettisoned, the gunner had to be ready to leave immediately after he initiated the turret jettison sequence.

The gunner was much more comfortably located in the forward compartment in the G and H models, away from the shake, rattle and roll of the tail compartment, and with his fellow crew members to draw moral support from. The upward ejection seat was also reassuring.

The fire control system in the G model was the AGS-15, operating the same four ·50 calibre M-3 machine-guns as the earlier mode. In the H, there was a significant increase in fire-power, with the installation of a six barrel M-61 cannon and the ASG-21 fire control system.

Oddly enough, in the raids over Hanoi, the crews preferred the old style D model arrangement, for the tail gunner could visibly track SAMs being fired from the rear, a tremendous advantage in the signal cluttered skies.

Munitions Handling

The ground crews of every air force are always the unsung heroes; the aircrew, whose work is admittedly more dangerous, always seem to attract the attention of news men, film makers, novelists and even their comrades in arms. But it is the people on the ground, mechanics, technicians, cooks and bakers, the whole enormous support network which ultimately keeps the force working.

No one group has been more overlooked than the dedicated troops who handle munitions. Their's is a dangerous task, the more so because they must repeat it so often, under the worst possible weather conditions, so that boredom and inattention become real problems.

Munitions handling probably reached its peak of efficiency—and hard work—during the Vietnamese conflict, where it was necessary for teams to work around the clock to keep the cavernous bomb bays filled.

SAC considers the task so important that it holds annual Munitions Loading/Security Police competitions, called "Giant Sword" which are as hotly contested as the bombing/navigation competitions are by flight crews. Weapons loading teams load both conventional and nuclear weapons aboard B-52 and FB-111 aircraft, with points awarded for all phases of the process, including use of handling equipment, and written tests on equipment knowledge.

The competition is important for morale reasons, but also because the intense preparation for it brings forth new and better weapons handling techniques each year.

Left: *The Mk 43 aerial bomb uses a variety of noses, and can either be parachute retarded or fall free. It also has a laydown capability (delayed surface burst) when dropped from altitudes above 200 feet.* (Chuck Hansen)

Below: *The four machine-gun rear armament was twenty years old before its actual combat debut; many had thought that it would not be useful. Two MiG 21s were shot down, three others were claimed, but the most useful service of the tail gunner was monitoring SAMs.* (Boeing)

FLYING THE BUFF

The B-52 has had trouble inspiring affection; it is a large aircraft, and the U in BUFF has always stood for ugly, even if the last F did not always stand for fellow. It demands much from its crews, not only because a mission is an exhausting combination of low level flight, inflight refuelling, demanding bombing runs and interminable navigation legs, but also because the environment is so bad.

Despite the great bulk of the fuselage, very little room is provided for the crew, who are dressed almost as uncomfortably as are astronauts. The cabin temperature varies from freezing cold to intensely hot, depending upon the time of day, position of the sun and inclination of the man in charge of the temperature controls. The pressurized air is bone dry, as is the oxygen the crew breathes when wearing masks, so sinuses become dry and irritated. The changes in pressure from surges in the pressurization system, changes in altitude and so on do not help. Add to this a high ambient noise level, the constant hum of electronic equipment, the static of radios and intercom and the intensity of concentration required for most jobs and you have a prescription for fatigue.

The aircraft is also giving constant signals as to the state of its health and its requirements for maintenance. There is nothing more boring and exasperating for an exhausted crew than the post-flight maintenance briefing. After an enervating 18 hour day, the crew must come back and fill out endless forms and answer probing questions from the maintenance staff. Yet if they do not recall every symptom sent to them—an errant vibration when engaging the autopilot, a transient signal in the radar set, a persistently low hydraulic pressure reading, a momentary failure of an alternator, a suspiciously low oil pressure reading—the aircraft might not be ready for its next take-off time, and the Wing Commander is very fussy about things like that.

The B-52 is really more than an aircraft, or even than a weapon system. It is a miniature universe, self contained, with millions of parts and six good men to use them over a long and complex mission. It has idiosyncracies peculiar to itself and to the fact that it was designed 30 years ago when some of the more sophisticated nuances of jet and swept wing aerodynamics were not known.

In the next few pages the experiences of pilots intimately familiar with the aircraft are told in their own words.

James H. Goodell has spent over half his life working at Boeing, and he is perhaps better qualified than any other person to talk about how the B-52 flies. Goodell flew patrol aircraft during World War II—PBMs and PB4Ys—and he got his Bachelor of Science degree in Mechanical Engineering in 1949. He tried the airlines for a while, didn't like it, and joined Boeing in February 1951, flying B-47s. He continued his test work on B-47 and B-52 for 11 years before transferring to Seattle, where he has flown 707s, 727s and 747s. He plans to fly until he reaches Boeing's mandatory retirement age from flying of 60.

Goodell writes "My general impression of the B-52 is that it is a good performing airplane, but has some operational drawbacks. The range, load carrying capabilities and cruise Mach No (Mach ·8) are outstanding, considering the era of the design. The control system and the handling qualities leave a lot to be desired. I am not sure what design considerations dictated the control system, but I do know that it caused some problems.

The airplane has a 10% chord elevator and rudder, making them smaller and less powerful than on most airplanes,

Below left: *A staged shot, showing the typical dash for the aircraft on an alert mission. Oddly enough, most missions flown in the B-52 have just the opposite feeling; the crew moves slowly and methodically about, doing their pre-flight duties with precision and never in haste.* (U.S. Air Force)

Below: *Navigators must still get intimate with the bombs, to be sure that the pins are properly pulled. Lieutenant James Wren hugs a 750 pound bomb on a B-52D's Multiple Ejector Rack (MER) at U Tapao.* (Major Dwight Moore)

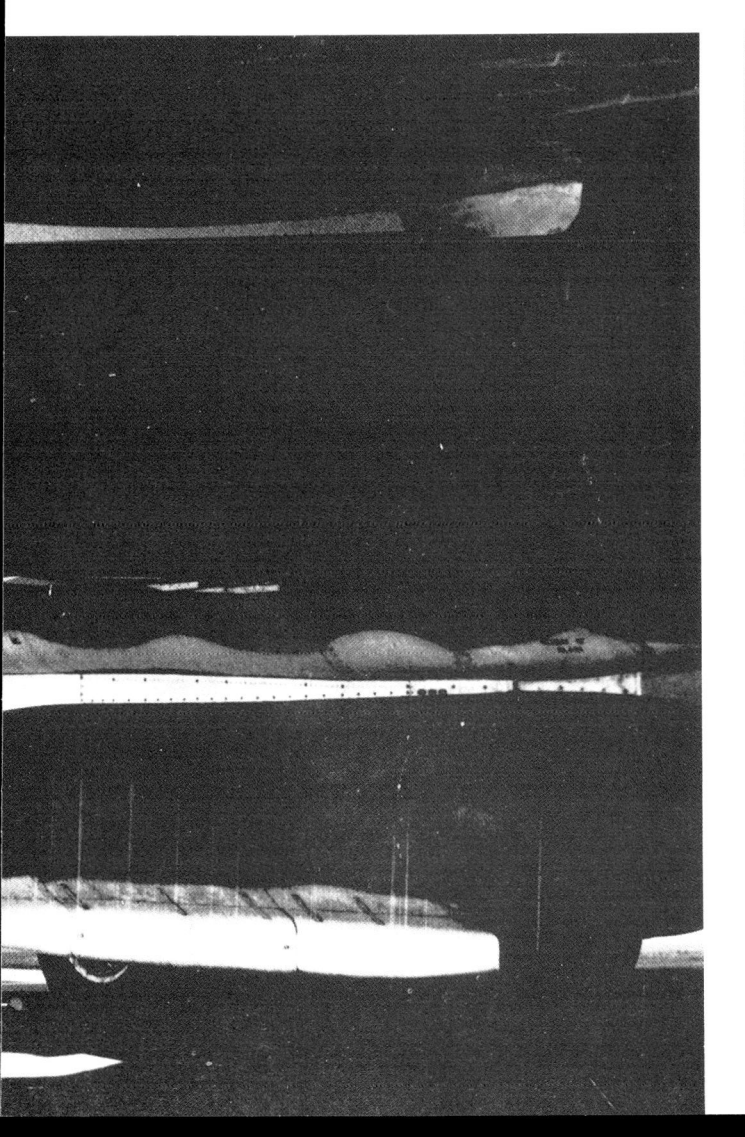

which typically have about 25% chord surfaces. The smaller elevator meant that the airplane could not be controlled in pitch with the elevator alone under all conditions. The stabilizer trim had to be used more than on other airplanes. Not all pilots remember this.

The narrow chord rudder meant that the airplane had to have a crosswind landing gear, because the rudder was not powerful enough to take the airplane out of a crab just prior to landing on a crosswind landing. The elevator and rudder controls were aerodynamic rather than hydraulic powered. (On early aircraft—*Author*.) This is probably why the smaller rudder and elevator were necessary.

There were some big pitch changes when operating the speed brakes. The airplane pitched up considerably when the speed brakes were raised. This had to be anticipated. The pilot could not hold the desired airplane attitude with elevator alone when the speed brakes were raised or lowered, but had to use a considerable amount of stabilizer trim. The same was true when the flaps were extended or retracted. The airplane pitched over considerably during the last part of flap retraction. In fact, a presentation was once made to B-52 pilots entitled 'No Distraction During Flap Retraction'.

The airplane is flown at altitudes of 200 feet up to 50,000 feet. It was not designed to fly at low level, but was made capable of doing so. It was not comfortable flying at low level in turbulence; such flight was described by one pilot as 'flying a bunch of loose parts in formation'. It did shake a lot.

I once flew on a test programme during which we flew several flights at an average altitude of 300 feet from Southern Texas to North Dakota, at speeds of from 325 to 400 knots. We ran into bugs, birds and turbulence. Everyone on board breathed a sigh of relief when each low level flight was over because we had an intuitive feeling that it was not safe. The navigator and bombardier were particularly upset with their situation as they sat on downward ejection seats. Of course at low altitude, the chances of anyone getting out were pretty slim in the case of structural failure.

Refuelling was a chore to be learned. Many SAC flights required a double refuelling, one after take-off and climb-out, and one in the middle of the flight. Most, but not all, pilots could learn refuelling. Under good conditions it took 15 or 20 minutes to take on 100,000 pounds of fuel. The goal was to do it without disconnects. When the pilot could do that, he was considered proficient.''

For a younger, and more contemporary point of view, Captain Michael Mastromichalis takes us on a typical practice mission. Mike is a B-52 aircraft commander stationed at Robins Air Force Base, Georgia, and he has this to say about a BUFF mission: "The day before we fly, the crew, which consists of six, meets at 8:00 AM in the squadron. During the day we prepare the flight plan, fuel-log, take-off data and other paper work that is needed to file a flight plan with the FAA. Our mission planning consists of about a five hour day which ends with a briefing . . . . . Once we are through the briefing we go into 'crew rest' which consists of 12 hours of uninterrupted rest prior to flying. Our take-off time is 10:00 AM, and the two pilots meet two and one half hours prior to that ('station time'—*Author*) at base operations. Here we check the weather and ensure that our flight plan has been accepted by the FAA computer.

After leaving base ops we report to the airplane one and one-half hours prior to take-off. The B-52 is a very complex weapons system and also older than a lot of the men who fly it. The tail number for my mission is 58-0212, so the airplane is about 20 years old. We have a 160,000 pound fuel load and no weapons.

Maintenance informs me that our hydraulic system needs to be fixed and there are several other malfunctions. About 20 maintenance men are scurrying around the aircraft, working to beat the take-off time.

We have an exact take-off time that must be met in order to accomplish the various segments of the mission, and although the B-52 is old, it very seldom has problems which prohibit flight.

At 15 minutes prior to take-off we start engines and begin to taxi, still with men working on the system malfunctions within the plane; they will be let out prior to take-off.

At exactly 10:00 AM we are ready for take-off; clearance is received and I apply power to all eight engines. When passing through 5,000 feet altitude, the radar observer calls that we have a thunderstorm dead ahead. I turn to clear the thunderstorm, then resume the planned course and level off at 28,000 feet with 370 knots indicated airspeed.

Once established at high altitude and constant airspeed, the B-52 is a very stable bombing platform. Our first activity will be refuelling with a KC-135A tanker.

We come into the air refuelling initial point, and the tanker is observed in a left-hand orbit. We are at 31,000 feet altitude now, and he is at 30,000. By controlling time—the time we are supposed to take on fuel—he will have rolled out ahead of me. We descend below him and close in to one mile range, then start a climb to close in to refuelling position, right under the tanker.

The actual flying of a B-52 during air refuelling is not easy. We have to keep the aircraft stable in a somewhat box shaped area so that the boom operator can put the end of the boom in our refuelling receptacle, which is just aft of the cockpit. Any deviation from the box shaped area (the 'refuelling envelope'—*Author*) and the boom will retract and disconnect so that it will not be broken.

The air refuelling lasts for about 25 minutes, of which we have the boom engaged for 17½ minutes, and receive 5,000 gallons of fuel. When we are through, I am drenched with sweat and my arms feel like lead. It is a period of intense concentration due to the danger of bringing two large aircraft into such close proximity at such high speed.

Next we start a celestial navigation leg. This is a one and one half hour event during which the navigator uses a sextant to 'shoot the sun' and plot the aircraft position. He uses time, heading and airspeed to arrive at the next position. Even if all the sophisticated onboard equipment were lost— inertial navigation, doppler, nav radios, etc—he would still be able to navigate anywhere in the world. His accuracy standards are plus or minus ten miles, at the end of the run, and he usually does much better than that.

After the navigation comes the 'meat' of the mission, the low level work. Throughout the US there are defined low

Above: *Boeing Test Pilot James Goodell doing a little inflight refuelling with, oddly enough, his gear down.* (Boeing)

level routes which are set aside for SAC's use. The one we will use is #41/42, which enters at Montgomery, Alabama, swings north and terminates in Georgia. We spend from two to three hours at low level—tiring work.

The low level starts by descending from high altitude down to about 400 feet from the ground. The first 15 minutes are spent evaluating the terrain avoidance radar to determine its reliability. At 400 feet above the ground at 320 knots indicated airspeed, at night, I want to be sure the system is perfect.

Once checked out, we trust the electronically generated trace which depicts the terrain ahead of us. By staying above the trace we can avoid the obstacles; of course we have to stay within the corridors established for the route, because it is airspace protected by the FAA (Federal Aviation Authority). The low level is tiring, and crew members often get sick from the rough air. Its no disgrace, after years of practice, for the hot air, bumps and tension make it miserable.

After about an hour and a half, we start the bomb run. Here a ground site manned by Air Force personnel can electronically score our bomb dropping capability. While the radar navigator uses his equipment to prepare to drop the simulated bomb, the Electronic Warfare Officer is receiving simulated anti-aircraft and missile attacks on his equipment. The crew has to coordinate well to fight off the attacks and still make an accurate bomb run.

Once the bomb run has been completed we can return to the base for some precious proficiency flying, which means practising a couple of instrument approaches and shooting two landings. After the last landing we taxi in, tired, but ready for the Maintenance debriefing, for the aircraft performed perfectly, and there are no write-ups.''

The final pilot report is from Captain Robert P. Jacober, Jr who flew 64 combat sorties, has 2099·9 hours of B-52 time and was SAC's first First Lieutenant aircraft commander. He is a second generation SAC man, for his father spent 26 years in the USAF.

Captain Jacober flew one tour in Southeast Asia as an aircraft commander, but he chooses to tell us of his first trip to Hanoi.

''My crew consisted of a veteran B-52 radar navigator, recycled C-123 'Ranch Hand' driver as the pilot, and a brand new navigator, gunner, EWO and me—the co-pilot. When we first arrived in Guam, in the summer of 1972, we were told the war was almost over. We probably would not fly the required 20 missions to qualify for an Air Medal. Then, each month, regular as clockwork, there was the 15th of the month rumour that the G models were going home. Every month that is, until December. For over a day, all bombing missions had been cancelled. Then, there was an aircraft commander's meeting. When the a/c returned from the briefing all he told us was that we should order two lunches, and to get lots of sleep. The rumours really started flying. War with China. War with Russia. G models going home. Everyone going home. Airborne alert for 24 hours.

All of the B-52 crews on Guam were assembled in the 'D' Complex (Arc Light) briefing room. Reminiscent of 'Twelve O'Clock High' episodes, the briefing officer stepped to the centre of the stage and said 'Gentlemen, your mission for today is . . .' A pause. Then the viewgraph machine projected an image of a small portion of a map on the screen. The map scale was such that 'HA' was on one side of a city and 'NOI' was on the other. The mind did not fuse the words until the briefing officer said 'Hanoi'. Dead silence was followed by everyone talking at once. Dramatic and impressive, yes. Scared, yes. Eager, yes. As a combat wing, this was the highest I had ever seen morale. We were about to really contribute to the war.

Again like on 'Twelve O'Clock High' everyone on base who could watched the aircraft take off. And everyone counted the planes when they returned, wondering where the missing were.

For this mission you preflighted a little bit more thoroughly, followed the checklist a little more closely, imagining what it would be like, and would you panic. Everyone was offered extra ·38 ammo, only our gunner took more. He took two extra boxes. We still don't know what he was going to do with all that ammo.

Crossing the pond westbound was the usual boredom. But spirits were high. There was some kidding. The gunner was bragging how he was going to get a MiG. The return trip was going to be more subdued, introspective.

My first real sense of combat was not the flashes ahead, that would prove to be the SAMs detonating, but the sound of multiple 'beepers'. These are the emergency locator radio beacons that are activated by a parachute opening after an ejection. So many going off at once could only signify that a BUFF had been shot down. The only advice I could remember was 'If you can keep a SAM moving across your windscreen, it is not going to hit you.'

We were lead, as the co-pilot, my primary jobs during

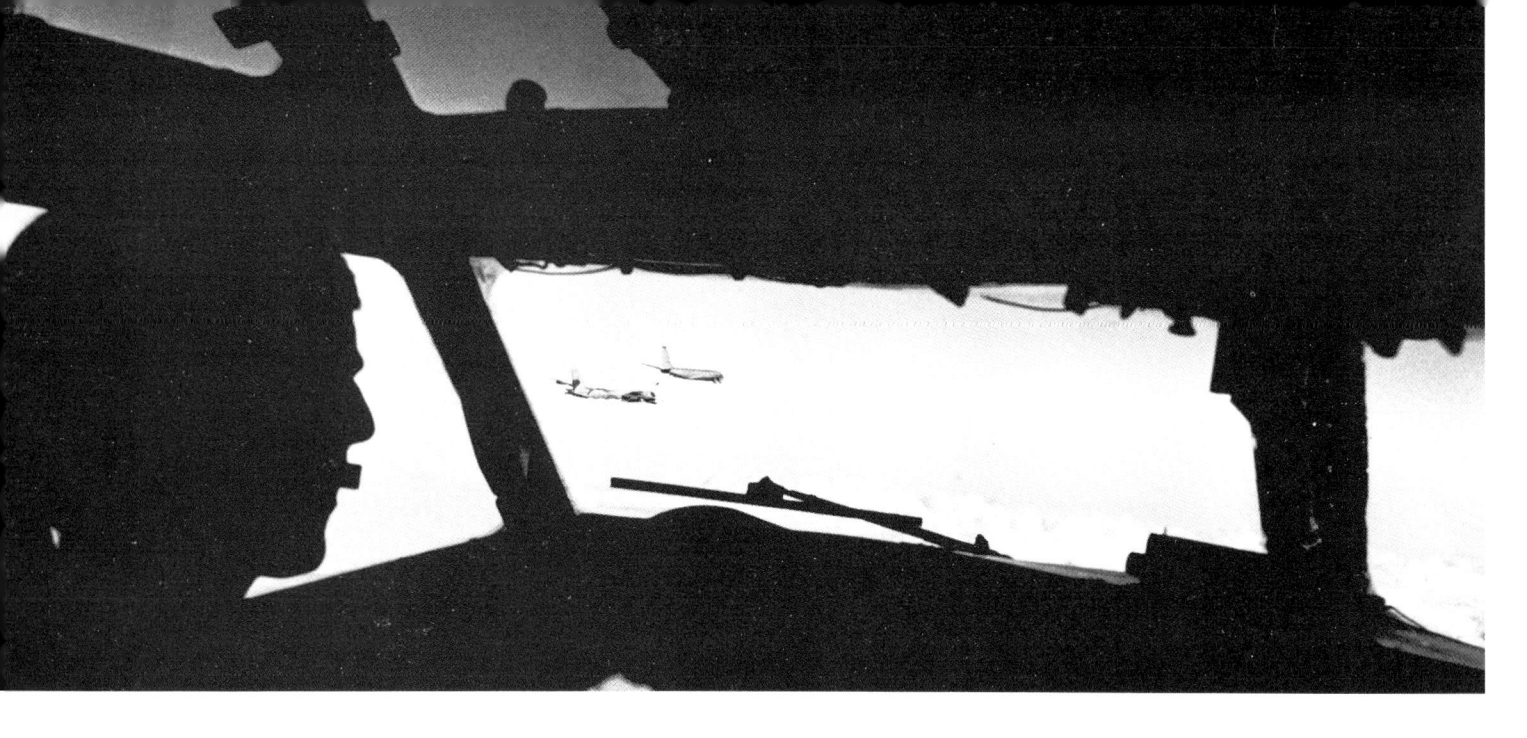

Above: A photo of the heaviest Boeing B-52 refuelling to date—Operation "Big Whip", in which the BUFF was loaded to over 550,000 pounds. (Boeing)

bombruns were: all communications between our aircraft and the other two aircraft in our cell; keeping the heading marker updated to the bombrun heading during the evasive manoeuvres to give the pilot a reference mark to roll out on, and keeping my head outside the cockpit to watch for anything. We had already been warned that MiGs had launched and were heading our way. The Electronic Warfare Officer (EWO) alerted us that SAM radars were following us. There was a cloud layer beneath us, so when I saw the first SAM ignite on the launcher, the plume was diffused like a match seen through a fog. But it quickly focused into a sharp flame, and the SAM looked like someone was throwing a candle at us.

Seven were launched at us during the bomb run, four on our inbound run, and three over the city. We could determine that only one was guided. It came at us from our twelve o'clock position. The EWO said "Uplink" about the time I saw it come up through the clouds. No matter what the pilot did, we could not get it to move on the windscreen. It passed just off of our nose and exploded several thousand feet above us. If they had had our altitude right, we probably would have been hit.

The bomb run itself went as briefed. Our post target turn took us among the four SAM sites that were in downtown Hanoi. During the turn, the pilot had rolled into about 70 degrees of bank. We lost several thousand feet. Since I was looking almost straight down, I had a good view of the three SAMs that launched. Fortunately none came close, and the return flight to Guam was quiet.

How was the B-52 to fly in combat? In Southeast Asia, we had air superiority in South Vietnam, and nothing on the ground could reach us. Much of the time you could not see the ground and rarely did the post target turn or track allow you to see your bombs go off. Actual combat was remote. The four hour trip over and the four hour trip back were usually flown by a pilot, safety observer and navigator. The other three were usually asleep, playing cards, writing etc. For the hour or two that you were actually over South Vietnam, the only differences were that we were strapped in tighter. The most dangerous part of the mission was watching the last 1,000 feet of runway approach as you tried to get a 480,000 pound airplane in the air. With a crew that worked and lived well together, the missions were tolerable. The sunsets and sunrises were spectacular and the Philippine Islands were postcard pretty.

Physically, the B-52 was cold after several hours of high altitude flying. The seat cushion didn't cushion after several hours, and the cockpit was cramped and uncomfortable. After the ten hours of the average SEA mission, it was pure pleasure to shut down the last engine, unstrap and really stretch. Though the BUFF could not jink, break or dive for the ground like the F-4 or the Thud, it was ideal for the mission that it supported. The B-52G could go unrefuelled from Guam to the target. At the altitudes we flew, we couldn't be seen or heard. Aside from the actual damage we wrought the psychological advantages of the BUFF were enormous... She may have her drawbacks, and be uncomfortable to fly, but she is a stable instrument and bombing platform. I guess like any pilot with his first combat airplane, I have come to love that venerable old airframe. She may be older than some of the pilots who fly her today, but age has not diminished her."

And that is perhaps as fine a commentary on a combat aircraft as possible—older than the pilots who fly her, but still undiminished. Not many combat planes have had that said about them, and none have deserved it so much as the incomparable Boeing B-52, the BUFF.

Appendix 1: B-52 Model Numbers

Air Force Model Number	Boeing Model Number
XB-52	464-67
B-52A	464-201-0
B-52B	464-201-3
RB-52B (Reconnaissance)	464-201-3
RB-52B (Day and night photo, weather and electronic reconnaissance)	464-201-1
RB-52B (Bomber version)	464-201-4
RB-52C (Bomber version)	464-201-6
RB-52C (Reconnaissance)	464-201-6
B-52C	464-201-6
B-52D	464-201-7
B-52E	464-259
B-52F	464-260
B-52G	464-253
B-52H	464-261

Appendix 2: B-52 Major Model Differences

Model	Air Force Serial Number	Production Unit No.	Initial Delivery	Recon. Cap.	Fire Control System	Bomb-Nav System	Engines*	Take-off Thrust Wet / Dry	External Drop Tanks (Gal. each)	Gross** Weight (lbs)	Operating Wt. Empty (lbs)****	Fuel Capacity (Gal)	Mil. Load** Capability (lbs)	Unrefueled Radius N. M. (MIL-C-5011A)
XB-52	49-230	1 Prototype	April 1953	No	None	None	4 - J57-P-8 / 2 - J75	N/A 10200 / N/A 15000	1000	405,000		38,865		3070
YB-52	49-231	1 Prototype	March 1953	No	None	None	YJ57-P-3	N/A 8700	1000	405,000		38,865		3070
B-52A	52-001 thru 003	1-3	June 1954	No	A-3A	None	J57-P-1W	11000 10000	1000	420,000		37,550	14,000	3110
B-52B	52-004 thru 013 52-8710 thru 8716	4-13 14-20	Sept 1954 June 1955	Yes Yes	A-3A MD-5	MA-6A MA-6A	J57-P-1W or J57-P-1WA or J57-P-1WB	11000 10000 / 11000 10000 / 11400 10400	1000	420,000	172,285 to 177,832	37,550	63,000	3110
	53-366 thru 376	21-31	Aug 1955	Yes	MD-5	MA-6A	J57-P-29W or	12100 10500						
	53-377 thru 391	32-46	Nov 1955	Yes	MD-5	MA-6A	J57-P-29WA or	12100 10500						
	53-392 thru 398	47-53	Feb 1956	Yes	A-3A	MA-6A	J57-P-19W	12100 10500						
B-52C	53-399 thru 408 54-2664 thru 2688	54-63 64-88	Apr 1956 June 1956	Yes Yes	A-3A (2688-MD-9)	AN/ASQ-48(V) AN/ASB-15 AN/APN-10b MD-1	J57-P-29WA or J57-P-19W	12100 10500 / 12100 10500	3000	450,000	172,637 to 179,390	41,550	64,000	3305
B-52D Seattle B-52D Wichita	55-068 thru 117 56-580 thru 630 55-049 thru 067 55-673 thru 680 56-657 thru 698	89-138 139-189 1-19 20-27 28-69	Oct 1956 Apr 1957 June 1956 Mar 1957 June 1957	No	MD-9	AN/ASQ-48(V) AN/ASB-15 AN/APN-10B MD-1	J57-P-29W or J57-P-19W	12100 10500 / 12100 10500	3000	450,000	170,126 to 180,811	41,550	64,000	3305
B-52E Seattle B-52E Wichita	56-631 thru 656 57-014 thru 029 56-699 thru 712 57-095 thru 138	190-215 216-231 70-83 84-127	Nov 1957 Feb 1958 Nov 1957 Dec 1957	No	MD-9	AN/ASQ-38(V) AN/ASB-4A AN/APN-89A MD-1, AJA-1	J57-P-29W or J57-P-29WA or J57-P-19W	12100 10500 / 12100 10500 / 12100 10500	3000	450,000	172,720 to 177,690	41,550	65,000	3320
B-52F Seattle B-52F Wichita	57-030 thru 073 57-139 thru 183	232-275 128-172	May 1958 June 1958	No	MD-9	AN/ASQ-38(V) AN/ASB-4A AN/APN-89A MD-1, AJA-1	J57-P-43W or J57-P-43WA or J57-P-43WB	13750 11200 / 13750 11200 / 13750 11200	3000	450,000	170,158 to 174,065	41,550	65,000	3345
B-52G	57-6468 thru 6520 58-158 thru 258 59-2564 thru 2602	173-225 226-326 327-365	Oct 1958 July 1959 June 1960	No	ASG-15	AN/ASQ-38(V) AN/ASB-16 AN/APN-89A MD-1, AJA-1	J57-P-43WB	13750 11200	700 (Fixed)	488,000	158,737 to 172,066	47,975	104,900	3785
B-52H	60-001 and on	366 and on	May 1961	No	ASG-21	AN/ASQ-38(V) AN/ASB-9A AN/APN-89A MD-1, AJN-8 J-4	TF33-P-3	N/A 17000	700 (fixed)	488,000	165,988 to 175,685	47,975 48,030***	105,200	4510

NOTES:
* -29W engines have 5,000 #/Hr water rate capability. These engines to be modified to 10,000 #/hr capability, making them -29WA engines.
  -19W engines are similar to -29W engines, but have titanium N₁ rotors.
  -43WA engines have 40,000 #/hr water rate capability and beefed-up accessory drive gears.
  -43WB engines have 40,000 #/hr water rate capability and beefed-up accessory drive gears, but are "flat-rated" on water.
** Weights shown are typical.
*** G and H with ECP 1050 wing.
**** Weight-range as of 25 March 1963

145

Appendix 3: History of B-52 Development

Model	Final Assembly First In	Final Assembly Last Out	First Flights	First Acceptance	First USAF Fly-Away*	Last Delivery**
YB			4-15-52			
XB			10-2-52			
A	7-20-53	6-28-54	8-5-54 (1)	6-17-54 (1)	11-27-57 (3)	6-25-59
RB	2-12-54	7-12-55	1-25-55 (5)	9-3-54 (7)	3-3-55 (5)	11-3-55
B	2-11-55	12-15-55	7-7-55 (28)	11-4-55 (28)	11-9-55 (28)	8-31-56
C	10-13-55	7-9-56	3-9-56 (55)	2-28-56 (54)	6-14-56 (55)	12-22-56
D(S)	5-28-56	8-7-57	9-28-56 (89)	11-30-56 (95)	12-1-56 (95)	11-1-57
D(W)	5-16-55	8-1-57	5-14-56 (1)	6-26-56 (1)	6-26-56 (1)	11-9-57
E(S)	7-1-57	3-10-58	10-3-57 (190)	10-7-57 (190)	10-7-57 (190)	7-2-58
E(W)	7-16-57	2-12-58	10-17-57 (70 & 71)	11-27-57 (73)	12-3-57 (71)	5-28-58
F(S)	1-20-58	12-3-58	5-6-58 (232)	6-18-58 (232)	6-18-58 (232)	2-25-59
F(W)	1-29-58	8-25-58	5-14-58 (128)	6-13-58 (128 & 129)	6-14-58 (128 & 129)	12-29-58
G	5-19-58	8-25-60	8-31-58 (176)	10-31-58 (173)	2-13-59 (183)	2-7-61
H	9-1-60	6-22-62	3-6-61 (371)	3-3-61 (368)	5-9-61 (366)	10-26-62

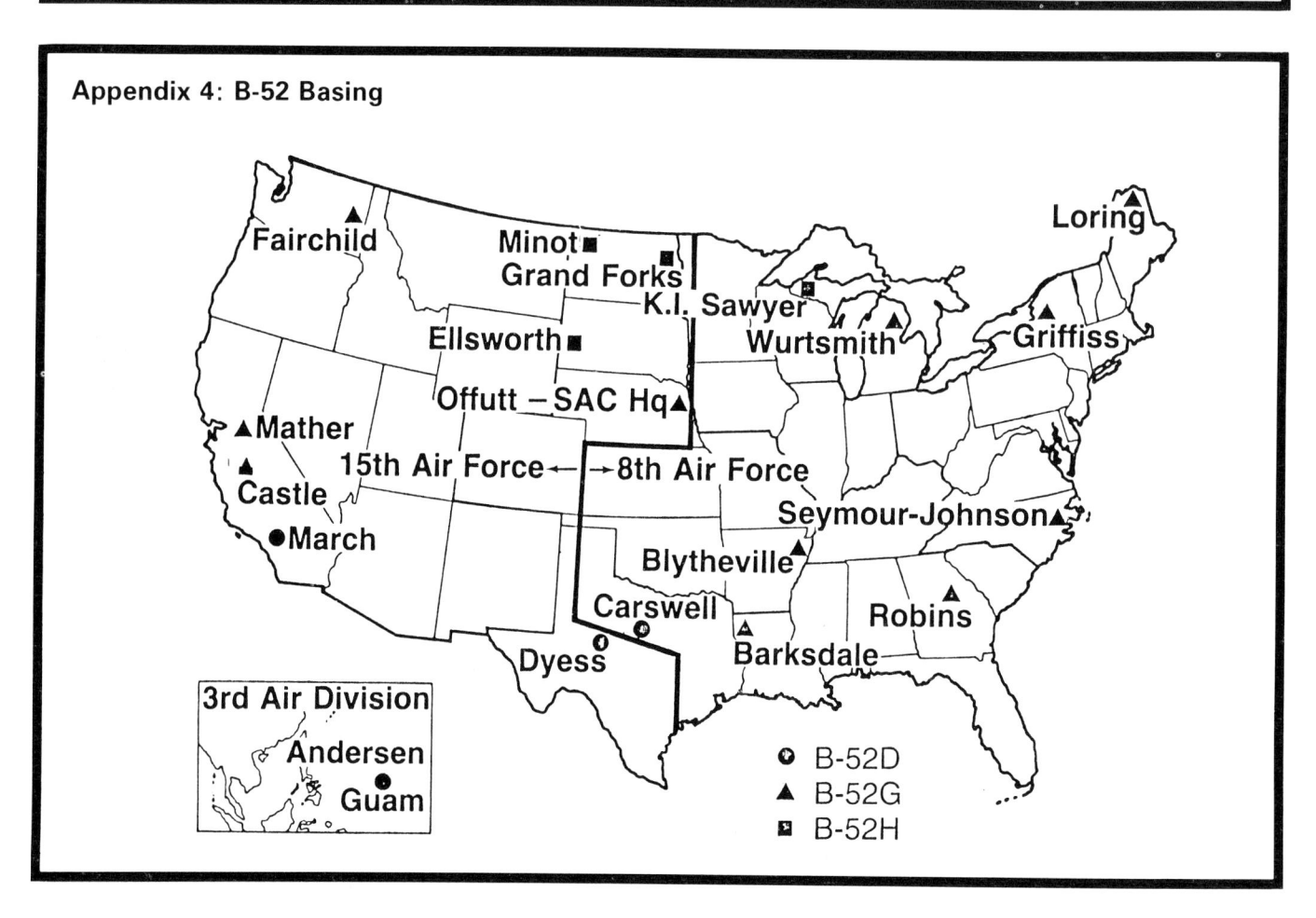

( ) = Seattle (S) or Wichita (W) Unit #   * = Excluding Boeing Test A/C   ** = Except Current Test A/C   Dates = Month-Day-Year

Appendix 4: B-52 Basing

Fairchild ▲, Minot ■, Grand Forks ■, K.I. Sawyer ■, Wurtsmith ▲, Loring ▲, Griffiss ▲, Ellsworth ■, Offutt – SAC Hq ▲, Mather ▲, Castle ▲, March ●, 15th Air Force ← → 8th Air Force, Seymour-Johnson ▲, Blytheville ▲, Carswell ◐, Dyess ◐, Barksdale ▲, Robins ▲

3rd Air Division: Andersen ●, Guam

◐ B-52D
▲ B-52G
■ B-52H

Appendix 5: B-52 Units

Unit	Base	Acft & Type	Date Models Assigned to Unit
93 BMW	Castle	B-52B	Jun 55-56
		B-52D	Jun 56-58; 65-74
		B-52E	1957-58; 67-78
		B-52F	1958-74
		B-52G	1966-67; 74-
		B-52H	1974-
42 BMW	Loring	B-52C	Jun 56-57
		B52D	1957-59
		B-52G	1959-
99 BMW	Westover	B-52C	Dec 56-71
		B-52D	1957-61; 66-73
		B-52B	1958-59
92 BMW/SAW	Fairchild	B-52D	Mar 57-71
		B-52C	1967-71
		B-52G	1970-
28 BMW	Ellsworth	B-52D	Jun 57-71
		B-52C	1967-71
		B-52G	1971-77
		B-52H	1977-
6 BMW/SAW	Walker	B-52E	Dec 57-67
11 BMW/SAW	Altus	B-52E	Jan 58-68
4123 SW	Clinton-Sherman	B-52E	Mar 59-Jan 63
70 BMW	Clinton-Sherman	B-52E	Feb 63-68
		B-52D	1968-69
		B-52C	1968-69
7 BMW	Carswell	B-52F	Jun 58-69
		B-52C	1969-71
		B-52D	1969-
4238 SW	Barksdale	B-52F	Aug 58-Apr 63
2 BMW	Barksdale	B-52F	Apr 63-65
		B-52G	1965-
4134 SW	Mather	B-52F	Oct 58-Jan 63
320 BMW	Mather	B-52F	Feb 63-68
		B-52G	1968-
4130 SW	Bergstrom	B-52D	Jan 59-64
340 BMW	Bergstrom	B-52D	1964-66
5 BMW	Travis	B-52G	Feb 59-Jul 68
	Minot	B-52H	Jul 68-
4228 SW	Columbus	B-52F	May 59-Jan 63
424 BMW	Columbus	B-52F	Feb 63-66
		B-52D	1966-69
		B-52C	1968-69
4138 SW	Turner	B-52D	Jul 59-Jan 63
484 BMW	Turner	B-52D	Feb 63-67
4241 SW	Seymour-Johnson	B-52G	Jul 59-Mar 63
68 BMW	Seymour-Johnson	B-52G	Apr 63-
4135 SW	Eglin	B-52G	Jul 59-Jan 63
39 BMW	Eglin	B-52G	Feb 63-65
72 BMW	Ramey	B-52G	Aug 59-70
97 BMW	Blytheville	B-52G	Jan 60-
4039 SW	Griffiss	B-52G	Jan 60-Jan 63
416 BMW	Griffiss	B-52G	Feb 63
4245 SW	Sheppard	B-52D	Jan 60-Jan 63
494 BMW	Sheppard	B-52D	Feb 63-66
4126 SW	Beale	B-52G	Jan 60-Jan 63

Unit	Base	Acft & Type	Date Models Assigned to Unit
456 BMW/SAW	Beale	B-52G	Feb 63-75
4128 SW	Amarillo	B-52D	Feb 60-Jan 63
461 BMW	Amarillo	B-52D	Feb 63-68
		B-52C	1967-1968
4038 SW	Dow	B-52G	May 60-Jan 63
397 BMW	Dow	B-52G	Feb 63-68
4043 SW	Wright-Patterson	B-52E	Jun 60-Jan 63
17 BMW	Wright-Patterson	B-52E	Feb 63-68
		B-52H	1968-75
	Beale	B-52G	1975-76
4170 SW	Larson	B-52D	Jul 60-Jan 63
462 BMW	Larson	B-52D	Feb 63-66
4137 SW	Robbins	B-52G	Aug 60-Jan 63
465 BMW	Robbins	B-52G	Feb 63-Jul 68
4141 SW	Glasgow	B-52D	Feb 61-Jan 63
91 BMW	Glasgow	B-52D	Feb 63-68
		B-52C	1967-68
379 BMW	Wurtsmith	B-52H	May 61-77
		B-52G	1977-
4136 SW	Minot	B-52H	Jul 61-Jan 63
450 BMW	Minot	B-52H	Feb 63-Jul 68
4042 SW	K. I. Sawyer	B-52H	Aug 61-Jan 63
410 BMW	K. I. Sawyer	B-52H	Feb 63-
4047 SW	McCoy	B-52D	Aug 61-Mar 63
306 BMW	McCoy	B-52D	Apr 63-73
		B-52C	1967-71
4239 SW	Kinchloe	B-52H	Nov 61-Jan 63
449 BMW	Kinchloe	B-52H	Feb 63-1977
19 BMW	Homestead	B-52H	Feb 62-1968
	Robbins	B-52G	Jul 68-
4133 SW	Grand Forks	B-52H	Apr 62-Jan 63
319 BMW	Grand Forks	B-52H	Feb 63-
22 BMW	March	B-52B	Sep 63-66
		B-52D	1966-
		B-52C	1967-71
		B-52E	1968-70
96 BMW/SAW	Dyess	B-52E	Dec 63-70
		B-52C	1969-71
		B-52D	1969-
4133 BMW(p)	Andersen	B-52D	Feb 66-Mar 70
43 SW	Andersen	B-52D	Apr 70-
72 SW	Andersen	B-52G	Jun 72-Nov 73
509 BMW	Pease	B-52D	Mar 66-69
		B-52C	1966-69
380 BMW/SAW	Plattsburgh	B-52G	Jun 66-71
4258 SW	U-Tapao	B-52D	Apr 67-Mar 70
307 SW	U-Tapao	B-52D	Apr 70-Sep 75
310 SW(P)	U-Tapao	B-52D	Jun 72-Jun 74
4252 SW	Kadena	B-52D	Jan 68-Mar 70
376 SW	Kadena	B-52D	Apr 70-Sep 70

Source: Command Historian SAC, Colonel John T. Bohn

Appendix 6: Capability Comparisons

	B-52D	B-52G	B-52H
Range – N Mi (High Altitude, Internal Payload)	6,400	7,300	8,800
Speed – KTAS	530	550	545
Combat Ceiling – Feet	45,000	46,000	47,000
Offensive Weapons – Max Quantity			
Gravity – Conventional	108	27	27
Gravity – Nuclear	4	8	8
Missile (SRAM) – Nuclear	0	20	20
Active Defense Weapons – Quantity and Type	4 .50 Cal Guns MD-9 FCS	4 .50 Cal Guns ASG-15 FCS	1 20mm Gun ASG-21 FCS
Passive Defense System – Quantity and Type			
Chaff	1125 Pkgs ALE-27	1125 Pkgs ALE-24	1125 Pkgs ALE-24
Flares	96 ALE-20	192 ALE-20	192 ALE-20
ECM – Type	Phase V	Phase VI	Phase VI
Low Level Capability – Feet	500	300	300
Terrain Avoidance Radar – Type	ASQ-48	ASQ-151	ASQ-151

Appendix 7: B-52 Production

Series	Seattle	Wichita	Total
XB-52	1	0	1
YB-52	1	0	1
B-52A	3	0	3
B-52B	50	0	50
B-52C	35	0	35
B-52D	101	69	170
B-52E	42	58	100
B-52F	44	45	89
B-52G	0	193	193
B-52H	0	102	102
	277	467	744

Appendix 8: B-52 Flyaway Costs

(airplane + spares + other (AGE))
(does not include "Peculiar Support")
1.5 to 1.9M per unit

Date of Contact	Number of Aircraft	Total Cost	Unit Cost
12/52	3-A; 17-RB	$ 594.5M	$ 29.5M
6/53	23-B; 10-RB; 35-C	495.4M	7.3M
9/54	50-D	226.7M	4.5M
12/54	27-D	277.0M	10.0M
	(Wichita, second source start-up)		
12/55	51-D; 26-E	313.7M	4.1M
3/56	42-D; 14-E	252.2M	4.6M
8/56	16-E; 44-F	246.4M	4.1M
8/56	44-E; 45-F	377.5M	4.2M
6/58	53-G	508.5M	9.6M
6/58	101-G	382.7M	3.8M
6/59	39-G	134.2M	3.4M
6/60	62-H	425.2M	6.8M
6/61	40-H	216.2M	5.4M
	744 Units	Total $4,511.0M	
		Avg. 6.1/Unit	

Appendix 9: B-52 Fleet Summary

Model	SAC Inventory	Test Inventory	Scrapped/ Attrited	Extended Storage	Total Built
B-52A	0	0	2	1	3
B-52B	0	1	49	0	50
B-52C	0	0	5	30	35
B-52D	79	0	44	47	170
B-52E	0	2	50	48	100
B-52F	0	0	30	59	89
B-52G	169	4	20	0	193
B-52H	96	0	6	0	102
	344	7	206	185	742

Appendix 10: B-52D

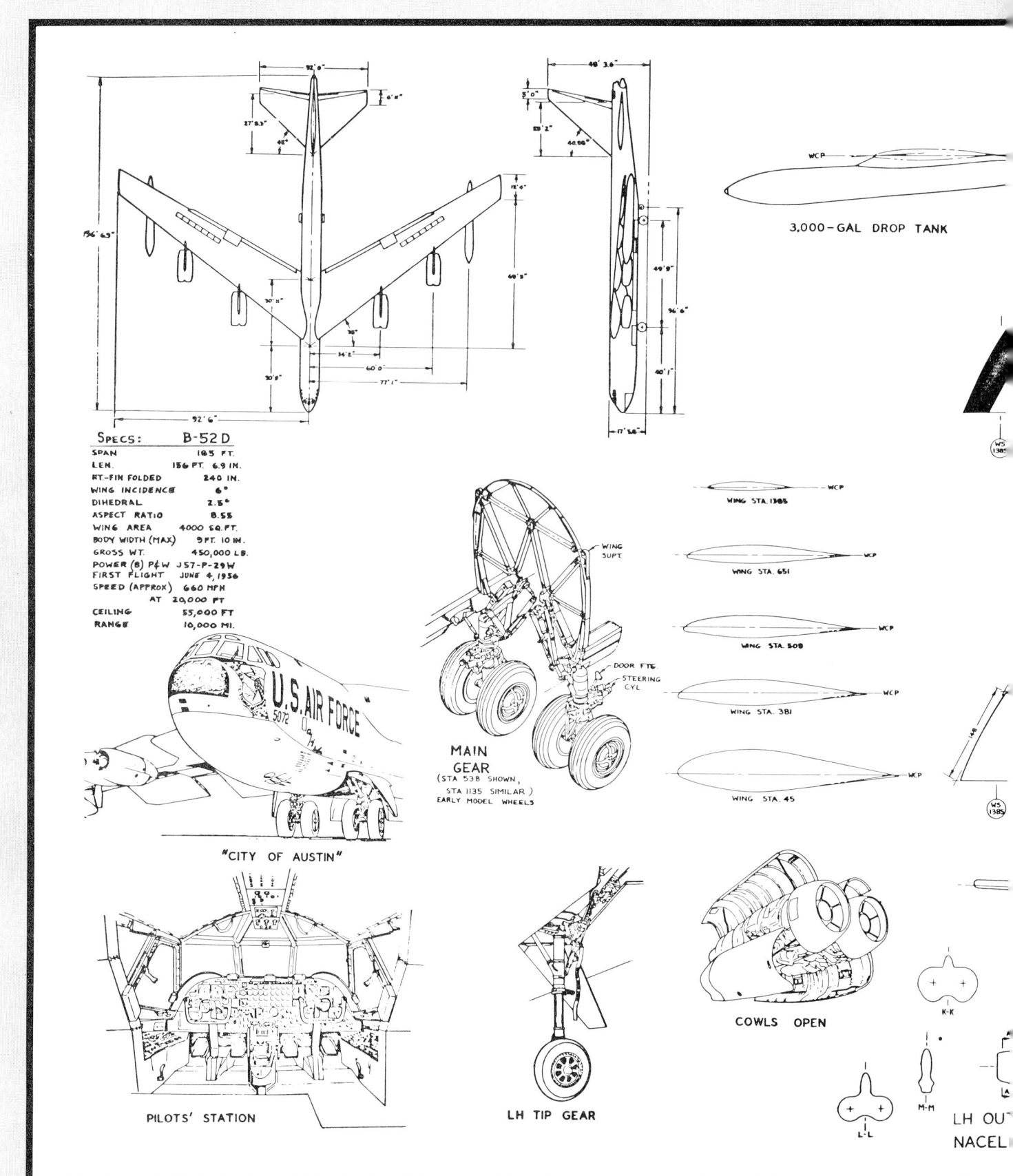

Specs: B-52D

SPAN	185 FT.
LEN.	156 FT. 6.9 IN.
RT-FIN FOLDED	240 IN.
WING INCIDENCE	6°
DIHEDRAL	2.5°
ASPECT RATIO	8.55
WING AREA	4000 SQ.FT.
BODY WIDTH (MAX)	9 FT. 10 IN.
GROSS WT.	450,000 LB.
POWER (8)	P&W J57-P-29W
FIRST FLIGHT	JUNE 4, 1956
SPEED (APPROX)	660 MPH AT 20,000 FT
CEILING	55,000 FT
RANGE	10,000 MI.

"CITY OF AUSTIN"

MAIN GEAR (STA 538 SHOWN, STA 1135 SIMILAR) EARLY MODEL WHEELS

3,000-GAL DROP TANK

WING STA. 1305
WING STA. 651
WING STA. 508
WING STA. 381
WING STA. 45

PILOTS' STATION

LH TIP GEAR

COWLS OPEN

LH OUT NACEL

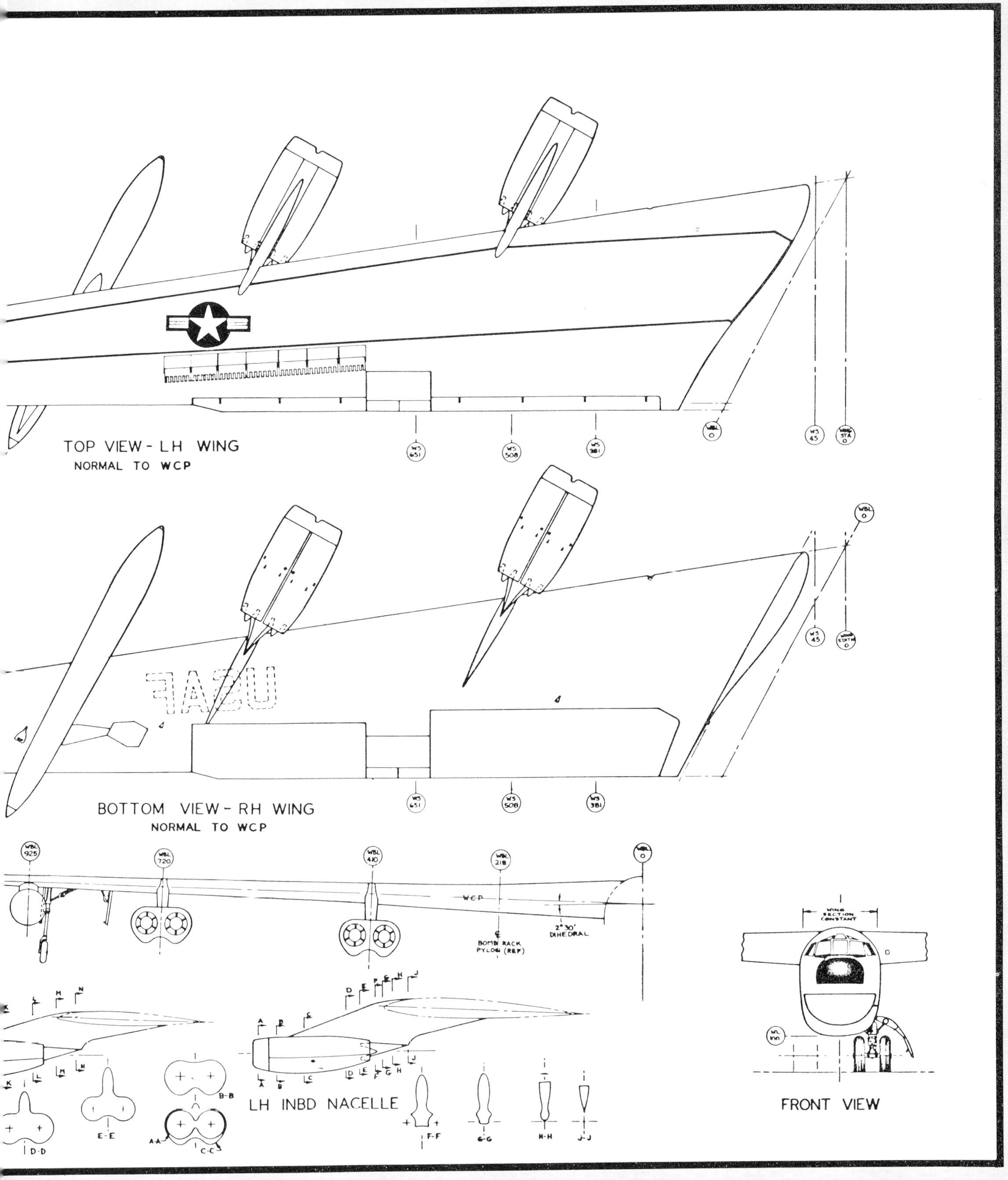
TOP VIEW - LH WING
NORMAL TO WCP
WS 651
WS 508
WS 381
WBL 0
WS 45
WING STA 0
BOTTOM VIEW - RH WING
NORMAL TO WCP
USAF
WBL 925
WBL 720
WBL 410
WBL 218
WCP
2° 30′ DIHEDRAL
℄ BOMB RACK PYLON (REF)
LH INBD NACELLE
D-D
E-E
B-B
A-A
C-C
F-F
G-G
H-H
J-J
WING SECTION CONSTANT
FRONT VIEW

FACTORY COLOR SCHEME

WHITE — UNDERSIDE OF WING, HORIZ. STAB. & FUSELAGE
  ENTIRE NACELLE & STRUT (EXC AS NOTED)
  ENTIRE DROP TANK
GRAY — AFT THIRD OF NACELLES, TAIL CONE, EXHAUST FAIRING
BLACK — ANTI-GLARE PANEL, WALK-WAY STRIPES
STRATA
 -BLUE — ALL LETTERING
DARK
 -BROWN — UPPER NOSE RADOME, WING TIP LE & UPPER SURFACE
   LE & TE OF FIN ANTENNAE
SILVER — UPPER SURFACE OF WING, HORIZ. STAB. & FUSELAGE
CLAD -
 ALUMINUM — VARIOUS UPPER SURFACE PANELS — SEE PHOTOS.
FIBERGLASS,
 -GREENISH — BASIC FIN ANTENNAE

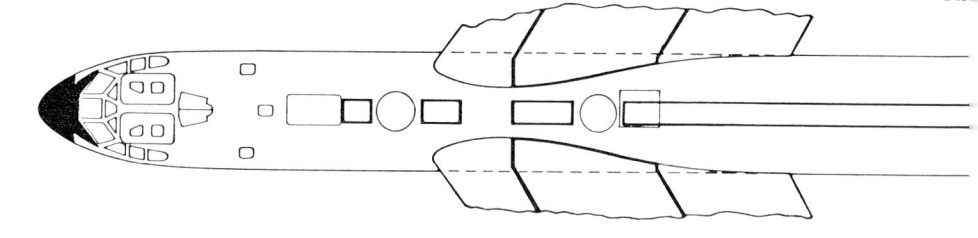

PLAN VIEW

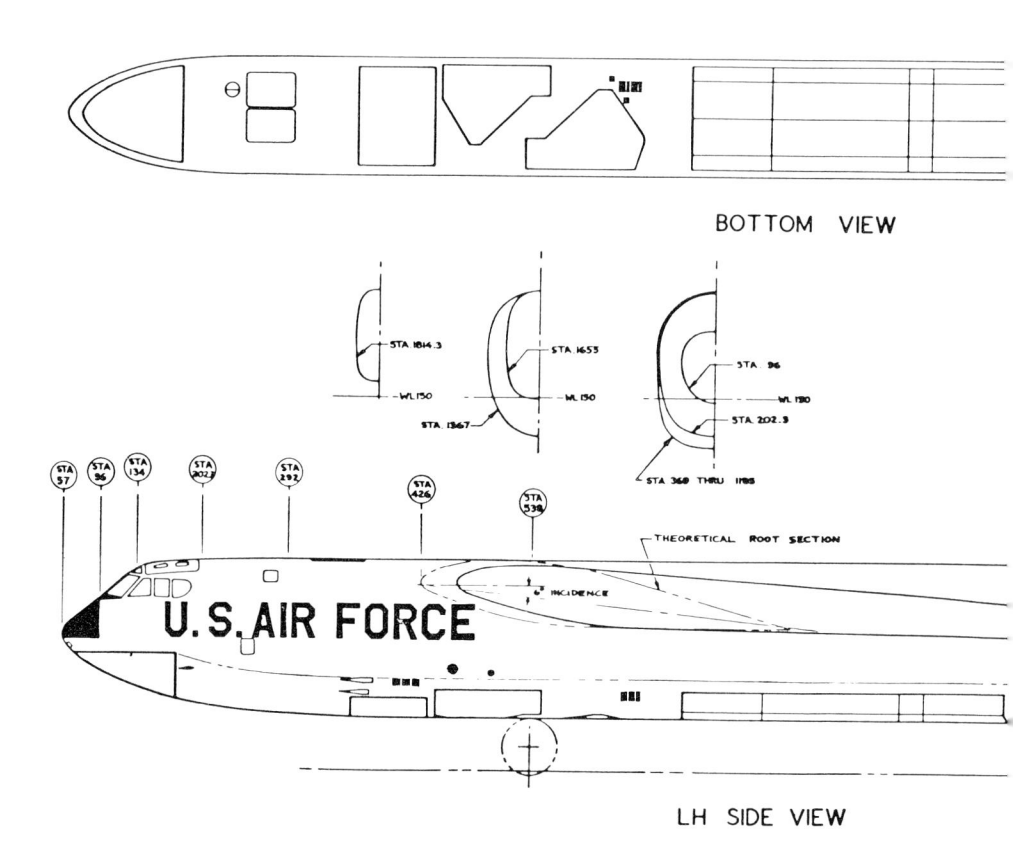

BOTTOM VIEW

LH SIDE VIEW

PROFILE

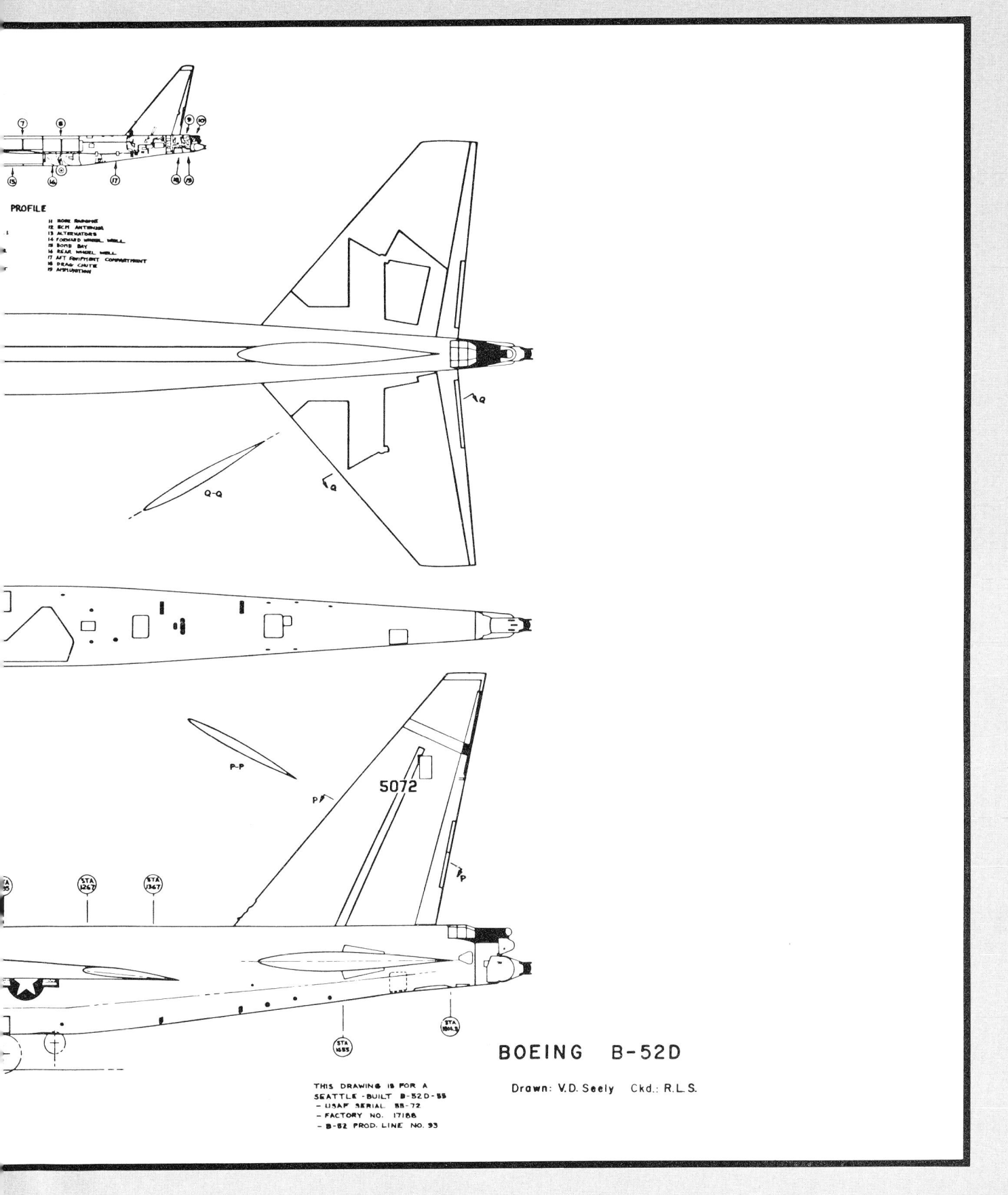

11 NOSE RADOME
12 ECM ANTENNAS
13 ALTERNATORS
14 FORWARD WHEEL WELL
15 BOMB BAY
16 REAR WHEEL WELL
17 AFT EQUIPMENT COMPARTMENT
18 DRAG CHUTE
19 AMPLIDYNE

BOEING B-52D

Drawn: V.D. Seely Ckd.: R.L.S.

THIS DRAWING IS FOR A
SEATTLE-BUILT B-52D-55
- USAF SERIAL 55-72
- FACTORY NO. 17188
- B-52 PROD. LINE NO. 93

153

Appendix 11: List of Major Modifications (over $50 million)

Mod #	Title	A/C	Years (FY)	Cost (M)
1000	Low level capability To improve bomber penetration capability by flying at 500 feet altitude or below; Includes: Terrain Avoidance Radar (ACR), Improved Radar Altimeter, Increased Cooling Capacity, Equipment Mounting Provisions, Secondary Structural Improvements	C-H	59-69	$ 313.2
996	AGM-28 Hound Dog Increased nuclear weapon capability with supersonic dash capability and additional penetration aids.	G-H	60-65	129.4
2126	AGM-69A SRAM Improve penetration capability	G-H	71-77	417.0
N/A	Stability Augmentation System Improve Dutch Roll damping, reduce structural loads and improve controllability in turbulence Includes: High performance rudder and elevator actuation system; stability augmentation system electronics, deletion of electronic and magnetic yaw dampeners	G-H	69-72	82.4
2595	Electro-optical Viewing System (EVS) Provides bomb damage assessment capability Includes: Low Light TV and Forward Looking Infrared sensors Integration with AN/ASQ-38A TA Radar, Controls and Displays at P/CP and Nav stations	G-H	71-77	248.5
N/A	Phase II ECM ECM capability improvements	C-G	61-69	127.6
2519	Defensive Avionics System (Phase VI) Enhance ECM capability Includes: AN/ALQ-117 Countermeasures Sets; Additional AN/ALT-28 Countermeasure Transmitters, Additional AN/ALE-20 Flare Dispensers, AN/ALR-46 Warning Receiver AN/ALQ-122 SNOE	G-H	70-77	362.5
2784	AFSATCOM World-wide communication capability Includes: Primary control with printer and keyboard, antenna, receiver/transmitter and modem	G-H	76/83	108.7
1581	B-52D Wing Mod (PACER PLANK) Strengthen wing and fuselage	D	72/77	219.4
1050	New Long Life Wing Design	G-H	59/64	139.1
1128	Body, Empennage Modifications (Empennage)	G-H B-H	62/66	87.9 199.1
1185-5	Fuselage Life Extensions	G-H	64-70	65.2
2525	NEW (ALQ-122) Improved ECM	G-H	71-82	76.5
2973	Tail Warning System (TWS) Provides threat detection, missile and aircraft, automatically initiates countermeasures (flares, chaff)	G-H	78-85	122.3
2970	ECM Transmitter Update Improved ECM capability; Updates ALT-28 jammers, replaces ALT-6Bs with ALT-28/ALQ-155 PMS	G-H	77-83	81.2
2973	ECM Power Management Update Upgrades ALT-28 to ALQ-155 power managed ECM which automatically responds to prioritized threat signals from ALR-46 radar warning receiver	G-H	77-83	126.0
3023	Avionics Modernisation (OAS) Enhances bomb/nav system by replacing majority of AN/ASQ-38 system with digital equipment; nuclear hardening of launch capability	G-H	79-86	1,425.7
3022 u	Cruise Missile (ALCM) Provides missile launch capability	G-H	78-86	1,041.2
951	High Stress, I, II, III Strengthening of critical structural areas	B-G	62-64	62.9
1195	Structural Modification Complementary mods to SAS system	G-H	69-71	82.4
F18411B	B-52D OAS Supportability Improvement (DBNS) Replaces 1950 vintage system with state of art digital	D	78-82	149.1
N/A	Defensive Fire Control	D-H	79-83	102.2
F19611B	Modernisation of ASG-21 Fire Control System Updates electronics with modular solid state systems	H	82/83	56.9
F3041	ODs/FRODs Wing root strakelets on G for SALT II recognition requirements	G	81-86	95

Source: AFLC Historian, courtesy Mr. Thomas M. Brewer, Mr. Roger Cummings and Captain John Adkison

Appendix 12:

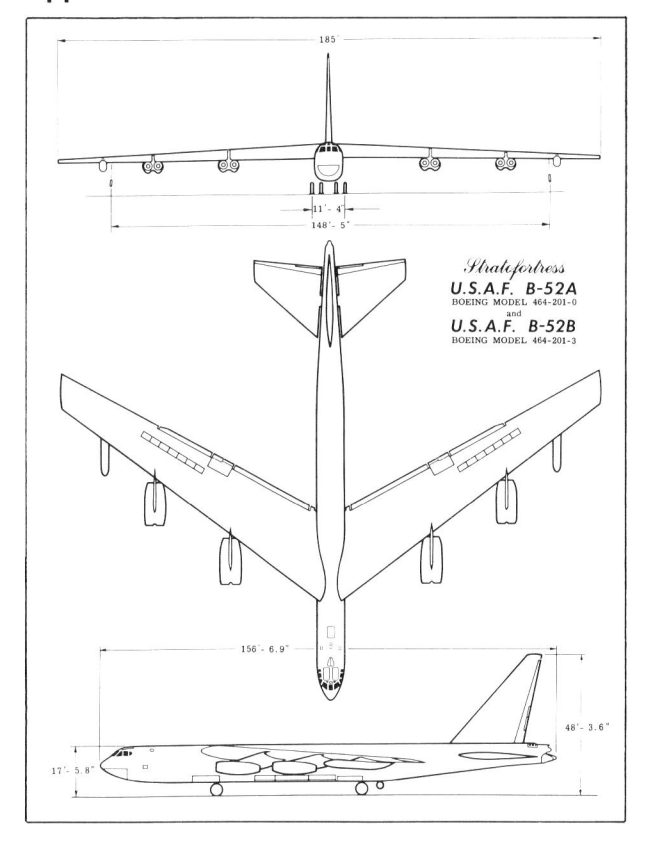

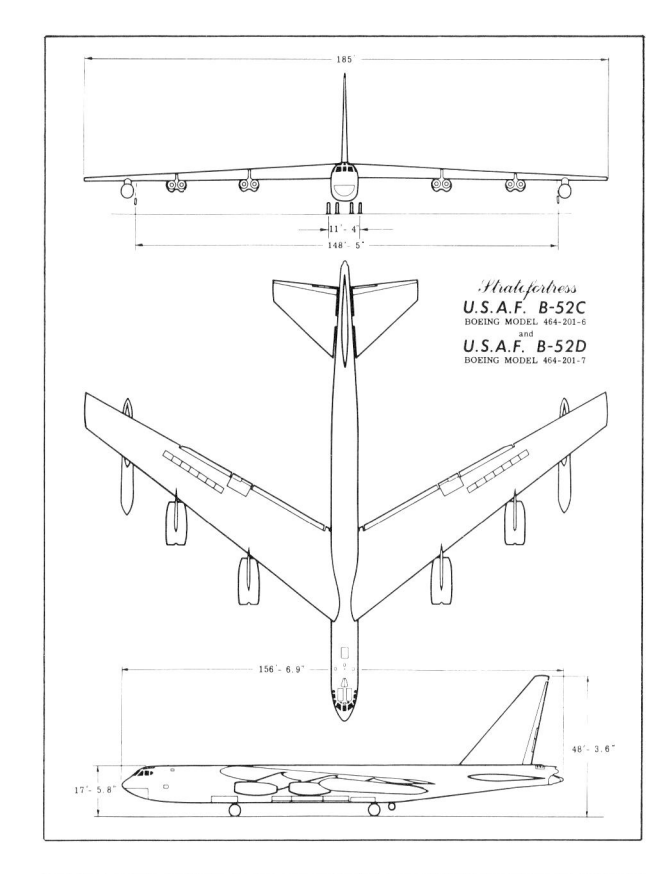

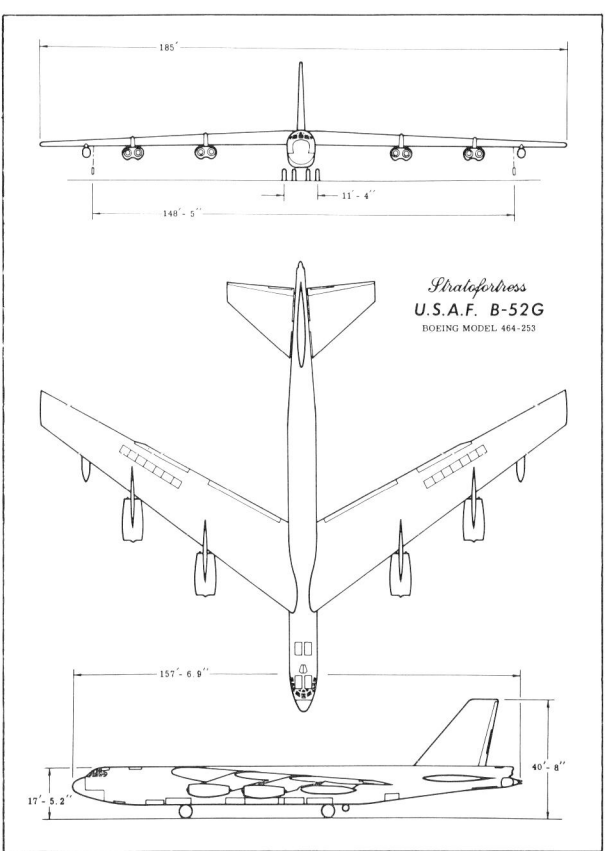

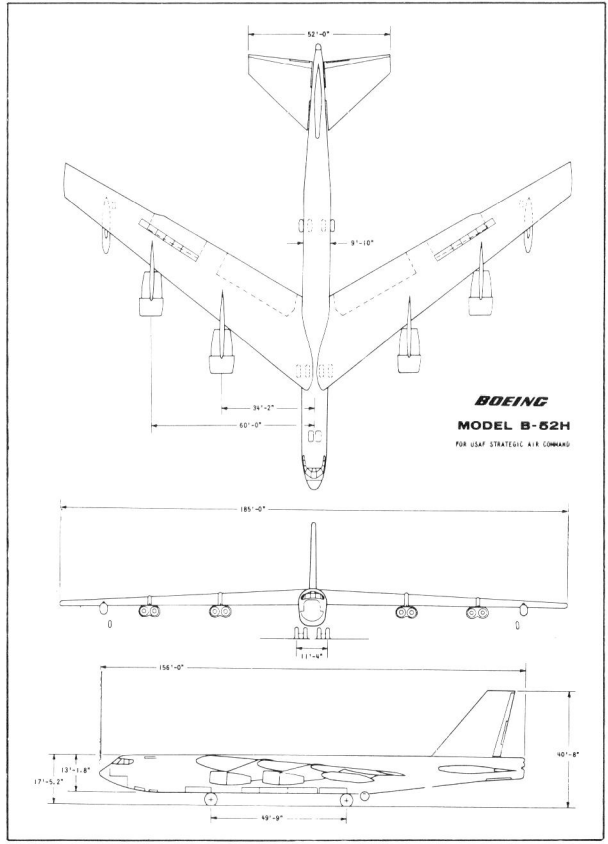

Appendix 13: Official Camouflage Paint Patterns

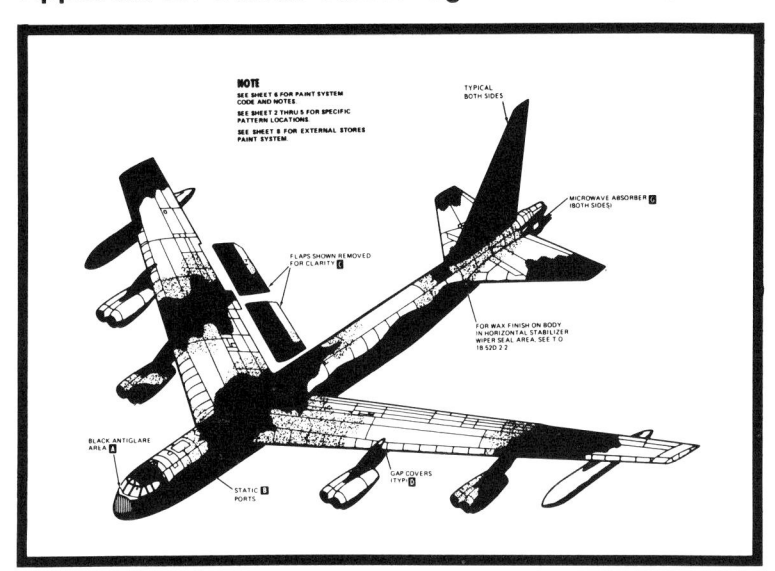

The official camouflage paint patterns for B-52D, G and H aircraft in tan, light and dark grey. The entire bottom of the D is of course black. (U.S. Air Force, Courtesy Boeing)

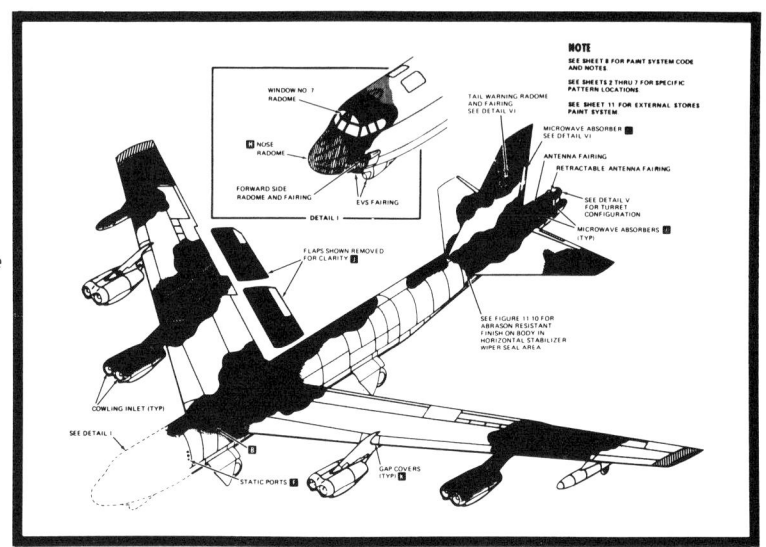

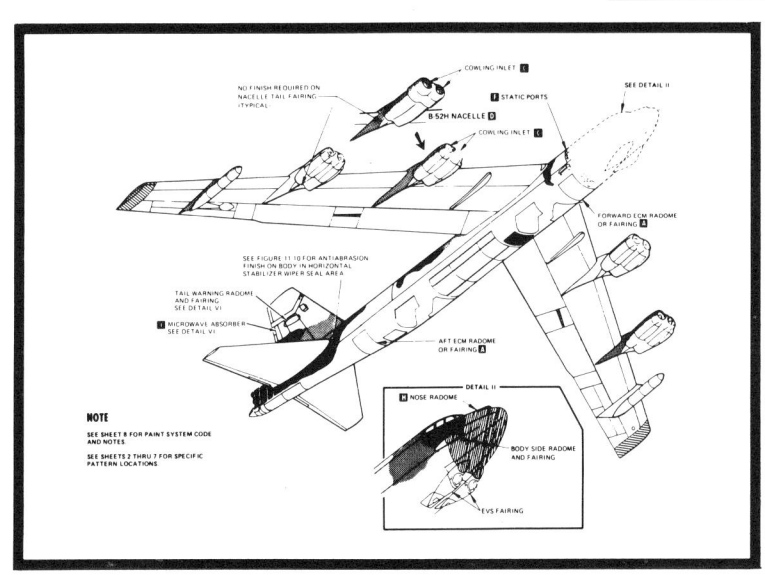

Appendix 14: Principal Personalities in the Development of the B-52

A series of biographical sketches of some of the principal players in the B-52 programme:

1 Edward C. Wells is now a consultant to the Boeing Company, completing his 50th year of service. He was born in Boise Idaho, on August 26, 1910, and was graduated from Stanford University with great distinction in 1931. He is a member of Phi Beta Kappa and Tau Beta Pi scholastic honorary fraternities.

Mr. Wells was employed by the Boeing Company in 1930, and had a meteoric career. He was first a design engineer, then became Chief of Preliminary Design in 1936, Chief Project Engineer-Military in 1938, Assistant Chief Engineer, in 1939, and Vice-President, engineering, in 1948. He was elected to the Board of Directors in 1951, and subsequently served as Vice President-General Manager of Systems Management in 1958, Vice President-General Manager of Military Aircraft Systems Division in 1961, and finally Vice President, Product Development in 1966. He became Senior Vice President-Technical until his retirement on January 1, 1972. Since that time he has served as a Director until 1978 and Company consultant.

His significant contributions to many Boeing programs, including the B-17, B-29, B-47, B-52, X-20, Minuteman, 707, 727, 737, 747, 757 and 767 have made him a legend in the Company. He has received many honors, including membership in the senior engineering associations, as well as honorary doctorates from the University of Portland and Willamette University.

Despite his incredible career and many awards, he remains a courteous, soft spoken man who continually deprecates his own contributions while signalling out others for their work. He is extremely precise in his choice of words, and has a tremendously broad overview of not only aviation and space, but also economics, and politics. Boeing was fortunate to have him, and in many ways he typifies the company—a doer rather than a talker.

2 George Swift Schairer is also a consultant to the Boeing Company, having retired in 1978 after 39 years of service. He began work with Bendix Aviation, moved to Consolidated Vultee, and then, in 1939, to Boeing. He has been Chief Aerodynamicist, Chief of Technical Staff, Chief Engineer, Director of Research, and finally, Vice President, Research and Development.

He received his Bachelor of Science degree from Swarthmore College, his Masters from Massachusetts Institute of Technology, and an honorary doctorate from Swarthmore.

Mr. Schairer has maintained a high degree of interest in education, and serves on the visiting committees of many universities, including MIT, UCLA, Stanford, Cal Tech and others.

During his service at Boeing he was a primary participant in many programs, including the B-47 and B-52, but one of his most signal contributions was his insistence on Boeing's building a large wind tunnel for its own use. The tunnel was of course extremely expensive, but it gave Boeing a technical lead during the 1940s and 1950s that it has never relinquished.

Like Wells, Schairer is a soft spoken, modest man, with a razor sharp brain. He too attributes to himself only the role of a "catalyst", but the warmth and admiration with which the other senior Boeing personnel speak of him belies this.

3 George C. Martin joined the Boeing Company in 1931 as a stress analyst. He later became Chief of Stress and was Staff Engineer for Structures and Power Plant.

In talking to Mr. Martin it is evident that he is proudest of his work as project engineer for the B-47 bomber which, as we have seen, laid the foundation for Boeing's later success in both military and commercial aircraft.

He was Chief Project Engineer during the design of the B-52, and at the same time worked closely with the Boeing 367-80 and 707 designs.

Mr. Martin became Chief Engineer for Seattle, and later was Vice President and General Manager of the Seattle Division. He subsequently became Vice President and Assistant General Manager of the Aero Space Division, and Vice President and General Manager for the Military Systems Division. For the ten years preceeding his retirement in 1972, he was Corporate Vice President for Engineering.

George Martin is an easy going, outspoken, humorous man whom it is a pleasure to interview and to know. Like his peers, he glosses over his own contributions, but expands on those of others. The spirit of cooperation he typifies is also typical of Boeing itself.

4 Holden White "Bob" Whittington is currently Vice President, Engineering of the Boeing Commercial Airplane Company. He graduated from MIT in 1941 with both a Bachelors and a Masters degree in Aero Engineering, and joined the Boeing company immediately to design and operate the company's transonic wind tunnel. He had a rapid rise through the Boeing system, culminating in his current position. He is a member of numerous professional societies, and has served on essential U.S. government advisory committees for the last twenty years.

5 Vaughn L. Blumenthal graduated from the University of Washington in 1940 with a degree in Chemical Engineering. He joined the Boeing Company in 1941, and in the following years was involved in aerodynamic performance analysis of the B-17, B-29, 377 Stratocruiser and C-97 aircraft. He was senior aerodynamicst on the B-52, and was then assigned to direct the B-52 Improvement Program which resulted in development of the F, and G models

In 1967 he was appointed Director of Product Development for the Commercial Airplane Division, and directed new and improved derivative airplane concepts for the 707 and 727 series. He is currently especially interested in aircraft noise abatement programmes.

6 Lawrence D. "Larry" Lee has been in aviation for the last forty years; his work has ranged from engineering test pilot to demonstration pilot to field service engineer. Presently the Customer Support Manager for the Boeing Military Aircraft Company in Wichita, Lee has assisted the Air Force in investigating most of the B-47 and B-52 accidents throughout the world. One of the most interesting and rewarding aspects of his job is participation in SAC's "Airborne Emergency Assistance Program"; he is the focal point for the "Conference Hotel" Telenet system for providing engineering technical data to aircrews having an emergency. An ebullient, genial man, Larry Lee is a pleasure to talk to; he has an unending fund of stories, some comic, some tragic, but all the stuff of aircraft novels.

7 Jack A. Nelson is the B-52 Program Manager at Boeing Wichita. An engineer with Boeing for more than 21 years, Jack has an encyclopedic knowledge of the B-52, gained during his years serving in a series of engineering and managerial positions. His personal flying tastes run to the zero engine rather than the eight-engine aircraft; he has an instructor pilot's license in gliders, as well as private and commercial powered pilot licenses.

8 Beverly W. "Bev" Hodges is a fast thinking, quick moving executive whose familiarity with the B-52's history is amazing. "Bev" galvanized the Wichita facility into four days of frantic action digging up old manuals, reports and photos for this book; much of the original material contained herein simply wouldn't have been available without his help. "Bev" retired after more than 38 years of service with Boeing, but he hasn't slowed down a bit.

Appendix 15: The B-52 Flies On

Right: *The B-52 of the future? An artist's impression of a four engine B-52, powered by the latest "jumbo jet derived" engines. It could happen, but because of the long development process, the decision has to be made soon if it is to be effective before the year 2000. (U.S. Air Force)*

Left: *As a part of the SALT negotiations, it was agreed to place "strakelets" on the wings of all B-52s modified to carry ALCMs. The agreement stipulated that the modification had to be visible from above, so that spy satellites could confirm the number of aircraft modified, and had to be made "aerodynamically and structurally integral" with the aircraft so that the change could not be quickly altered, or switched from aircraft to aircraft. As a result, air intakes were built into leading edge of strakelet. This is the test aircraft, B-52G 58-204.*

Left: *First flight of a B-52 with the XTF39 engine built by General Electric. It has approximately the same thrust as four of the B-52's J57 engines. (Air Force)*

Right: *Another proposal to take the B-52 into the 21st century features individually podded turbofans, chin ride-control surfaces and drag-reducing winglets. (Boeing)*

INDEX

Numbers in **bold type** refer to picture captions

A-7	98
AGM-86B	**132**
AGM-109 "Tomahawk"	**132**
Air Force units	
376th Air Refuelling Wing	98
330th Bomb Squadron	27, 37
7th Bomb Wing	91
22nd Bomb Wing	37
43rd Bomb Wing	37
93rd Bomb Wing	27, 37, 64, 67
96th Bomb Wing	96
100th Bomb Wing	39
306th Bomb Wing	37, 96
320th Bomb Wing	91
379th Bomb Wing	**84**
410th Bomb Wing	**80**
416th Bomb Wing	**116**
4017th Combat Crew Training Squadron	64
303rd Consolidated Maintenance Wing	96
43rd Strategic Wing	100
4136th Strategic Wing	80
4252nd Strategic Wing	90
See also Eighth AF, SAC, and Twentieth AF	
ALCM	132, **132, 133,** 135
Alexander, John	53
Allen, William	104
Allison engines	
J35	**31**
J35-2	35
American heavy bomber development	11-26
Arnold, Gen. Hap	**17**
Atlantic Charter	43
Atlas weapon system	105
Avro Vulcan Mark II	128, **128**
B-1	43, 53
B-2 Condor	13, **16**
B-10	**17,** 18, **20,** 21
B-17 Flying Fortress	11, 21, 22, **22, 23,** 26
success of	22-3
B-18	22
B-24 Liberator	23
B-29	23, **23, 24,** 26, 27, 44
B-32 Dominator	**24**
B-36 Peacemaker	**26,** 29, **31,** 44, **52**
B-45	29
B-47	11, 26, 27-9, **31,** 32, **35, 36, 38,** 50, 53
B–E models	36
characteristics	27
development and test	28-9, 32, 34-6
engines	34-5, 36
first flight	34
inflight refuelling	27, **35**
operational problems	39, 42
operational use	36-7, 39
stages of design	**32**
B-50	**24,** 26, 27, 44
B-52	39, 43-4, **56, 57, 60**
acceptance of	52-3
adaptations	9
Advanced Capability Radar in	80, 114, **115**
armament	55, 137-8, **139**
average units costs	63
bailout procedures	72-3
bomb capability	135-7, **141**
bomb load	**93**
Bomb/Nav system improvements	113-14
cockpit	**82**
conditions for birth of	49-50
crew compartments	**86-7,** 140
crew positions	**62**
crew "scramble"	**141**
Dayton conference on	50-2
defensive avionics systems	**120**
de-icing	55
drag chute deployment	**71**
electrical system	55
Electronic Counter Measures (ECM) equipment	116, 119, **119, 120**
Electro-Optical Viewing System (EVS)	114, **115,** 115-116
empennage	54-5
engines	55-6, **84**
(*see also* Pratt & Whitney engines)	
fuselage	54
hydraulic system	55
in action in Vietnam	**90, 91,** 91, 93, 94, **94,** 98
in-flight refuelling	142, **143,** 144
landing gear	55, **84**
lift to drag ratio	81, **84**
loading bombs on	**92, 93,** 135
long service	9, **103**
low level work	80, 142-3
manufacturing and maintenance centres for	68
missile armament	123, 125, 127-32, 135
munitions handling	138
non-stop world flight	**65,** 66-67
"nuclear hardening" test for	**122**
Offensive Avionics Systems	119-20, **120**
personality	10, 140, 141
pilot's instrument panel	74-5
planning, advent of	56
pneumatic system	55
pressure suits in use	**62**
production aircraft	63-7
Project "Pacer Plank" modifications	**107,** 107-8, **108-9, 111,** 111, **112,** 112-13
progression of design	**44-7**
rear gun turret	**69**
roll out and first flight	58-9, 62
servicing diagram	**88**
SRAM-carrying	**130, 131**
structural modifications	104-5, **106,** 107
systems displayed	**80**
tail "stinger"	**84**
wing	53-4, **55**
B-52A	**62,** 63
B-52B	64
B-52C	**67,** 67
B-52D	67, **68, 69, 70,** 93, 94
modifications for Vietnam	113, 114, **135**
shared systems with G and H	114
B-52E	68, **70,** 76
B-52F	**70,** 76, 113
B-52G	76, 77-8, **78, 83,** 96, **116,** 144
modifications	114, 119, 120
shared systems with D and H	114
with EVS and OAS	**122**
B-52H	78, **79,** 79-80, **81, 83,** 129
modifications	114, 119, 120
shared systems with D and G	114
B-58	43, 53, 95
B-70	53
Barling bomber	13, **16**
Beall, Wellwood E.	**57**
Bell, Lawrence D.	11, **12**
Blumenthal, Vaughn	50, 53, 59
Boeing Company	
and B-52 programme	103-4
Model 299	**20,** 21, 22
Model 424	29, **32**
Model 448	**32,** 34
Model 450-1-1	**32,** 34
Model 450-65-10C	**43**
Model 462	44
Model 464 variants	**44-7,** 51
Model 474	**49**
Model 479	**49**
wind tunnel	54
Bolling Air Base	89
Burchinal, Col David A.	37
C-123	143
C-130	98
Caproni Ca 5	11, **12**
Carlsen, Art	50, 53, 58
Castle Air Force Base	64, **65,** 90
Cavalry Division	
1st Air	93
Congress, and aircraft procurement	43
Curran, Art	**57**
Curry, Dick	113
Curtiss V-1570-29 Conqueror engine	18
Dingwell, Capt	67
Douglas, Donald W.	11, **12**

EA-3	98
EB-66	98
EC-121	98
Eighth AF	96, 98
Eisenhower, Pres. Dwight D	90
Elias XNBS-3	13, **13**
Elrod, Capt John	**57**
F-4	98, 144
attacked by SAM	**100**
F-84	105
F-86	28
F-100 Super Sabre	28
F-105	98, 144
FB-111	43, 95, 98
Felix, Dale	113
Fisher, Charles F. "Chuck"	**112**
brings in damaged B-52	112-13
Funk, Jack	114
GAM-72 Quail missile	123, **123**, 124, 128
GAM-77 Hound Dog	77, **125**, 127-8
GAM-87A Skybolt	77, 80, **127**, **128**, 128, 129, **129**
"Gatling Gun"	55, **79**
GBU-15 glide bomb	114, **135**
General Electric engines	
J47-GE-3	35
J47-GE-11	35
J47-GE-25	36, **38**, 44
J85	**124**, 125
Goodell, James H.	141, **143**
Grissom, Col T. R.	78
Grumman Panther	28
Guam	91, 93, 97
Anderson Air Force Base	**96**, 98, 100
bomb assembly area	92
missions from	143-4
Handley Page 0/400	11, **12**, 13
Hanoi, bombing of	98, **99**, **100**, 118, 144
Harris, Gen Harold R	**16**
HH-53	98
Higgins, Paul	59
Hill, Major Ployer	22
Hillman, Col Dan	67
Hobbs, Leonard S.	56
Hodges, Beverly W. "Bev"	103
Holloway, Gen Bruce K.	93, 116
Howell, Major Joseph W.	27
Huff-Daland	
LB-5A	**16**
XLB-5	13
Jacober, Capt Robert P.	**100**, **102**, 143
Jewett, Robert "Bob"	**31**
Johns Multiplane	13, **15**
Johnson, A. M. "Tex"	**57**, 59, **60**
Johnson, Lady Bird	62
Johnson, Lt Gen Gerald W.	96
Jones, Gen David "Davey"	**128**, 129
Jones, Robert T.	29, 32
KC-97	27, **35**
KC-135A tanker	68, 90, 98
Kennedy, Pres John F.	90, **128**, 129
Kenney, Gen George C.	89
Keystone bombers	13
Kindleberger, Dutch	11
Kirtland, 1st Lt Roy C.	9

Lee, Lawrence D.	97
LeMay, Maj Gen Curtis E.	26, 36, 39, 50, 54, **57**, 63, **67**
accepts B-52 concept	52
in command of SAC	89
Little, Helene K.	104
Loesch, Richards	**57**
LWF Owl	13, **15**
MA-2 autopilot	80
MA-2 bombing system	**76**
MB-1	13
MB-2	**12**, 13
McCarthy, Col James R.	100
McCoy, Col Michael N. W.	37
McDonnell, James	11
Macmillan, Harold	129
McNamara, Robert	129
Martin, George	**31**, 34, 53
Martin, Glenn L.	11, **12**
builds first bomber	13
forms company	11
Martin GMB	11, **12**, 13
Martin (J.V.) Cruising Bomber	13, **14**
Martin (J.V.) Kitten	**14**
Martin Model 146	21
Martin TT	11
Mastromichalis, Capt Michael	142
Me-262	32
Meyer, Gen John C.	95, 98
Mitchell, Gen Billy	13, **16**
More, AIC Albert	100
Morris, Lt Col James H.	67
Nazarro, Gen Joseph	93
NBS-1	**12**
Norden bombsight	19, 37
Nuclear bombs	135-6, **136**, 137, **137**, 139
Oklahoma City Air Material Area (now Logistics Center)	68, 104
Olds, Maj Gen Archie	66, **67**
P-26	**18**, 21
Page, George	**16**
Pennell, Maynard	50, 53
Polaris missile	129
Powers, Gen Thomas S.	90
Pratt & Whitney engines	55
development	55-6
J52-P-3	127
J-57 and variants	50, 51, 52, 56, 63, 64, **70**, **76**, 79
P-43W	76
PT series	56
R-1860-11 radial	**17**
R-1860-13 Hornet	18, 21
R-4360 "Corncob"	**24**, 44
TF33	56, **78**, 79
Putt, Lt Gen Donald	**60**
PV-2 Neptune	80
Rascal missile	**42**, 123
RB-52B	**64**, **65**, 104
Roche, Jean A.	19
Russian radar systems	68
Ryan, Gen John D.	90
San Antonio Air Material Area (now Logistics Center)	68, 104
Schairer, George S.	**31**, 32, 50, 52, 53

Schlech, Major Russ	27
See, John	116
Spaatz, Gen Tooey	18
Springer, Eric	**12**, 13
SRAM	129-30, **130**, **131**, 131-2
Stocker, Major Bill	97
Strategic Air Command (SAC)	36, 37, 39
actions in SE Asia	90, 91, 93-101
Arc Light mission in Viet Nam	**94**, 102
Bullet Shot operation	**94**, 96
"Charlie Tower" control innovation	97
crew units	102
equipped with B-52s	67
established	89
hostile penetration targetting	**117**
launch plan during Viet Nam war	**102**
Linebacker I mission	96
Linebacker II mission	**94**, 95, 96-97, 98, 100, 102
missile/aircraft nature of	89, 90, 95
training programme	89, 90, 93-4
Townsend, Major (later Brig Gen) Guy M.	34, **35**, 52, 55, **57**, 66
Turboprops	44
Turner, T/Sgt Sam	100
Twentieth AF	26
Viet Nam war	90, 91, 93-101
Warden, Col Henry E. "Pete"	34, 50, 51, 52, 53
Warner Robins Air Logistics Center	104
Wells, Edward C.	21, **31**, 50, 52, 53, 58
sketches of B-52	**50**
Willgoos, Andrew V. D.	56
Wilson, T. A.	58
Withington, H. W. "Bob"	50
Wolfe, Gen K. B.	34, 50
Wren, Lt James	**141**
Wright engines	
R-3350 Cyclone	**18**, 19, 26
T35 turboprop	44
YT49-W-1 turboprop	36, **40**
Wright, Orville	11
XB-9	13, 18
XB-15	**18**, 21, 43
XB-19	**19**, 21, 43
XB-35	**25**, **31**, 44
XB-36	44
XB-42 "Mixmaster"	**28**
XB-43	**28**
XB-45 Tornado	28, 29, **29**
XB-46	28, **31**
XB-47	29, **32**, **40**, 44, 49
XB-48	29, **31**
XB-49	**25**, **31**
XB-52	49, **50**, 52, 53, 58, 59, **60**, 62
XB-55	**49**, 49, 53
XB-70	43
XB-907	**16**, 18
XNBS-3	13, **13**
XNBS-4	**15**
YB-49	29
YB-52	56, **58**, 59, 62
YB-60	52, 63
Y1B-19	**17**
Y1B-26	**17**
YDB-47E	**42**